Quantum Inference and the Optimal Determination of Quantum States

K. R. W. Jones[1]

University of Bristol

H.H. Wills Physics Laboratory

Royal Fort, Tyndall Avenue,

Bristol, BS81TL

A thesis submitted for the degree of

Doctor of Philosophy

December 18, 1989

[1]Research supported by a Commonwealth Scholarship.

Memorandum

I declare that no part of this thesis has been submitted for a higher degree in this, or any other, University. The research reported herein is the result of my own investigation unless reference is made to the work of others. All research was carried out under the supervision of Dr. J. H. Hannay, at the University of Bristol, between October 1986 and October 1989.

Kingsley R.W. Jones
December 1989

Acknowledgements

I should like to thank my supervisor John Hannay for engaging my interest in what proved to be a very rewarding line of research. I am especially grateful to him for sharing the odd mathematical trick, for exposing me to deep questions and for his willingness to discuss strange ideas. I register some brief but particularly useful discussions with Asher Peres, who brought the No–Cloning Theorem to my attention. This proved significant in my thinking.

For financial support, so necessary to overseas study, I thank the Commonwealth Scholarship Commission for making research possible and the British Council for their comprehensive administration to all needs. Further to this I am grateful to my parents for contributing much needed support, of every kind, during writing.

Finally, I should like to thank the many other students I have interacted with at Bristol, both in the line of research and play. Singularly outside this group I express my sincere thanks to Bente for her support, general good thoughts and patience with frustrated long–range interactions.

Abstract

This work involves an exploration of the application of techniques from the theory of Bayesian Inference and from the theory of Communication Systems to the problem of Quantum Measurement. Such ideas are not new, they were pioneered by W.K. Wootters in his PhD thesis[1]. However, new application is made here to the problem of determining quantum states. In particular, we define and solve the problem of how to make an optimal determination of the state prepared by some piece of apparatus.

Initial attention focuses upon reviewing the established theory of Quantum Measurement with a view to highlighting two features.

Firstly, single states are in principle undeterminable. This possibility is defeated by the Quantum No–Cloning Theorem[2, 3]. This proves to be a relativistic requirement and emerges naturally from the formalism in a way that demonstrates the intricate symbiosis between the two quantum evolutionary processes, as formulated by von Neumann.

Secondly, emphasis is given to the view that the most fundamental requirement of measurement is an inference of the system state. Given the above observation it is necessary to have an ensemble of quantum systems available to derive useful information about their preparation. The historical focus has been upon the measurement of expectation values. There are well known difficulties with this in relation to the measurability of operators. An approach that is more naturally tailored to operational methodology involves treating the quantum probability law as a conditonal over a number of outcomes determined by the Hilbert space dimensionality. The possibilty of altering these probabilities through choice of different measurement operators then enables accumulation of sufficient statistical data to infer the state in a very natural way via Bayesian Inference. General operator expectations are then *calculable* from expectations upon projectors (measured probabilities).

Application of the above mentioned concepts allows a natural formulation of the quantum limits to information gain. This is made possible by the concept of Correlation Information which is naturally applicable to any problem of Bayesian Inference. An Optimal State Determination Problem is formulated and studied in detail for two state systems where the necessary measurement operators are fully realisable. General conclusions are made for Hilbert spaces of higher but finite dimensionality.

The resulting theory does not involve changes to the accepted paradigm. However, it does provide a picture of Quantum Measurement that is more naturally in line with the attempts at pre-axiomatisation made by Jauch[4] and others. Operational tests of the transformation laws of Quantum theory are indicated. These are independent of the Correspondence Principle. Ideas are then sketched for a possible new class of experiments and connection is made with recent work by Balian et al.[5, 6, 7] that is very similar in approach.

For Arthur and Wendy.

Contents

1 Quantum Measurement Theory   **1**

    1.1   Quantum Axioms   1

         1.1.1   Hilbert space, states and density matrices   3

         1.1.2   Dichotomy of Processes   5

         1.1.3   Conclusions for state determination   9

    1.2   The analyser-preparator couple   11

    1.3   Geometry of Density Matrices   13

2 Bayesian Inference   **20**

    2.1   A simple measurement model   20

    2.2   Bayes' Rule of Inference   22

    2.3   Properties of the Prior Distribution   24

    2.4   Examples   26

         2.4.1   Gaussian correlation   26

         2.4.2   Coin probabilities   27

3 Correlation Information, Communication and Measurement   **30**

    3.1   Information theory   30

    3.2   Communication and Correlation Information   32

    3.3   Communication and Measurement   35

    3.4   Properties of the

        Correlation Information   37

4 Formulation of the OSDP   **41**

    4.1   The Quantum Die   41

    4.2   Optimal State Determination   42

    4.3   Summary   46

i

5 Two state OSDP   **47**

5.1   Sphere formulation for two states 47

5.2   Simplification of the
       Correlation Information formula 49

5.3   Singlet probability formula 51

5.4   Triplet probability formula 52

5.5   General probability formula 55

5.6   Numerical experiments 61

       5.6.1   An initial guess: the Platonic solids 61

       5.6.2   Rationale for a computer search 61

       5.6.3   Calculation algorithm 62

       5.6.4   Optimisation algorithm 63

       5.6.5   Numerically generated optimal configurations 63

5.7   Asymptotic Correlation Information for
       repeated measurement sets 70

       5.7.1   Outline of calculation 70

       5.7.2   Calculation of G 75

       5.7.3   Observations concerning G 80

       5.7.4   Numerical tests of G 81

       5.7.5   Calculation of g 88

       5.7.6   Asymptotic formula 91

5.8   Uniform Measurement Set 95

5.9   Upper bound to the
       Correlation Information 101

5.10   The inferred density matrix 104

6 General OSDP   **108**

6.1   Ray integration measure 108

6.2   General measurement bases 111

6.3   Simplification of the
       Correlation Information formula 112

6.4   General probability formula 115

6.5   Mutually unbiased bases:
       numerical experiments 118

6.6   Uniform measurement set 119

6.7   Upper bound to the
       Correlation Information 127

7 Conclusion   **131**

     7.1   Conclusions about Quantum Inference 131

     7.2   Tests of Transformation Theory 134

A Asymptotic Information for Singlet measurement   **A-1**

B Best Geometries for small N   **B-1**

List of Figures

1.1 Geometry of state space 18

3.1 Discrete communication channel 33
3.2 Communication and measurement 36

5.1 Low N optimality 66
5.2 Variation with opening angle for the septet 68
5.3 Correlation Information: triplet and singlet 71
5.4 Plot of weight values 82
5.5 Plot of log prior probabilities 85
5.6 Steepest descent excluding correction for sphere curvature . . 86
5.7 Steepest descent including correction for sphere curvature . . . 87
5.8 Triplet anisotropy contours 96
5.9 Tetrahedron anisotropy contours 97
5.10 Dodecahedron anisotropy contours 98
5.11 Icosahedron anisotropy contours 99

7.1 Schematic of proposed experiment 138

List of Tables

5.1   Best configurations 65

6.1   Mutually unbiased bases $d = 3, 5$ 120

B.1   Irregular Quintet B-1

B.2   Irregular Septet B-2

B.3   Irregular Octet B-2

B.4   Tetrahedron vertex directions B-2

B.5   Dodecahedron face directions B-3

B.6   Icosahedron face directions B-3

Chapter 1

Quantum Measurement Theory

1.1 Quantum Axioms

The treatment of measurement theory will be based upon axiomatic quantum theory. This formalises the rules employed in quantum calculations and provides the foundation for discussions of quantum measurement.[1] It is well known that the resulting theory of measurements can be fully realised for certain two state systems, see [1]. For other systems many of the situations discussed may not admit experimental realisation. Such practical considerations will be ignored as it is the optimality constraints imposed by the formalism which are sought.

Familiarity with the following axioms is assumed. Properties of Hilbert space and the space of states are of special interest.

- The state of a quantum system is fully described by a ray in an associated Hilbert space. Conventionally this is denoted by a ket:

$$|\psi> \text{ or } |\psi(t)> .$$

Constraints may apply so that not all rays correspond to states[13, p.304]. However, all states correspond to rays.

- The time evolution of a quantum system is governed by two processes.[2]

---

[1] General references for this chapter include [4, 8, 9, 10]. Note added: there is a lucid account of these topics in chapter six of the recent book by Penrose[11].

[2] The terminology is due to von Neumann.

Which is chosen depends upon (defines) the degree of isolation of the quantum system.

Process One : The system is *not isolated* and makes a stochastic transition to $|i>$, one of an orthonormal set of possible new states $\{|i>\}_{i=1}^d$. This occurs with probability

$$\Pr(i) = <\psi|i><i|\psi> \ .$$

Such a process is called a complete measurement. The possible new states can be influenced by the manner in which this process is brought about. Normally, one considers them to be determined as the eigenvectors of some Hermitian operator $\hat{A}$. This represents an observable of the system. The eigenvectors $|i>$ of this operator determine the probabilities whilst the eigenvalues a_i (necessarily real) define an expectation value for the observable in virtue of the relations:

$$< a > = \sum a_i \times \Pr(i),$$

$$< a > = <\psi| \left[\sum a_i |i><i| \right] |\psi> = <\psi|\hat{A}|\psi> \ .$$

Expectation values are not of importance here since, as we shall show, it is only the probabilities that yield information about ψ.

Process Two : The system is *isolated* and evolution proceeds deterministically. The state changes continuously according to the *linear* equation:

$$i\hbar \frac{d}{dt} |\psi(t)> = \hat{H}(t)|\psi(t)>$$

with $\hat{H}(t)$ a self adjoint operator called the Hamiltonian.

Three aspects of the axiomatic system are important to us: the nature of Hilbert space and state space, the dichotomy between Processes One and Two and the quantum probability rule. The first two are discussed in the following subsections. Discussion of the dichotomy of processes leads naturally to an understanding of quantum limits to knowledge. Consequences of the quantum probability rule are the major concern of this thesis and so occupy centre stage later on.

2

1.1.1 Hilbert space, states and density matrices

Concerning the Hilbert space, a selection is made according to the nature of the quantum degrees of freedom. Continuous degrees of freedom select infinite dimensional Hilbert space. Generally these are separable and so all examples of the same cardinality are isomorphic[12]. Discrete degrees of freedom, peculiar to quantum mechanics, select a finite dimensional Hilbert space. All examples are isomorphic when of the same dimension[12]. This means that quantum systems with the same number of states d have identical state spaces. Only finite dimensional systems will be of interest here.

Loosely speaking, finite dimensional Hilbert spaces carry the properties of particles and their internal degrees of freedom. However, they may also describe any situation where there are a finite number of possible quantum outcomes.[3] In any case there is always an infinite dimensional Hilbert space (for the "motion") associated with any finite dimensional Hilbert space. This will allow us to demand that the finite part is free of constraints such as antisymmetrisation where several identical systems are involved.

The importance of the above-mentioned elementary isomorphism is that all d–state systems, will have the same state determination theory. This is because the state determination will proceed by probabilistic inference on Process One considered as a conditional probability rule:

$$\Pr(i|\psi) = <\psi|i><i|\psi> .$$

The possible probabilities for fixed ψ depend on the possible choices of measurement basis $\{|i>\}_{i=1}^{d}$. These are determined solely by the dimensionality d. So although electrons are spin half and photons are spin one, the masslessness of the latter imposes constraints such that it behaves as a two–state system[14] and the two can be dealt with by the same $d = 2$ state determination theory.

The actual space of quantum states is not the Hilbert space but the rays in Hilbert space. Kets must be normalised and are identical up to phase. The phase must be pulled out to get the space of states. The easiest way to do this is to form the operator

$$|\psi><\psi|.$$

---

[3]Examples: Dead−or−alive states of Schrödinger's cat, Wigner−Weisskopf atom and the Double−Slit experiment, see [14].

3

Taking normalised kets as ingredients and forming such operators, all kets belonging to the same ray map to one such operator. Mathematically speaking, they lie in equivalence classes and a single such operator tags each class. The operators are called pure-state density matrices. Often they are not really matrices ($d = \infty$) but for finite dimension they genuinely are.

Pure states are the legitimate states for single systems. It will be found necessary to deal with many copies of identically prepared systems. The state of such an ensemble is described by a general density matrix. This can be defined as a *Trace 1, non-negative Hermitian* matrix, denoted ρ. From this mathematical definition the physical meaning can be extracted.

The density matrix represents an enlargement of the notion of state. In an ensemble of identical systems, individual constituents may well have different states but there is no way to distinguish them. The density matrix packages information about the expectation values that can result upon observation of this ensemble. To see this note that any matrix of the above form admits the spectral decomposition

$$\rho = \sum_{i=1}^{d} \lambda_i |i><i|,$$

where

$$\sum_{i=1}^{d} \lambda_i = 1 \text{ and } \lambda_i \geq 0, \lambda_i \in \mathbf{R}.$$

Calculate the quantity

$$\mathrm{Tr}[A\rho] = \sum_{i=1}^{d} \lambda_i \mathrm{Tr}[A\rho_i],$$

$$\rho_i = |i><i|$$

where Tr denotes the matrix trace. Using

$$\mathrm{Tr}[A\rho_i] = <i|A|i>$$

it is clear that

$$\mathrm{Tr}[A\rho] = \sum_{i=1}^{d} \lambda_i <i|A|i> .$$

To interpret this note that the λ_i sum to one. So the quantity on the right, being the average of the expectation values on pure states $|i><i|$, can

be interpreted as the expectation value for ensemble measurements. The ensemble behaves like a mixture of systems in single system pure states each with weight λ_i. The above decomposition into pure states is not unique. It comes from diagonalising the density matrix. This canonical decomposition of the density matrix is useful but does not imply that the ensemble actually contains these pure states in the given proportions. Any convex combination of density matrices

$$\rho' = \lambda\rho_1 + (1 - \lambda)\rho_2, \ \lambda \geq 0,$$

is also a density matrix. This shows that many decompositions are possible. Linearity of the matrix trace ensures that the above interpretation of ensemble expectation is consistent.

The only occasion on which ρ admits a unique decomposition is when one eigenvalue is one and the others are all zero. It is then clearly *idempotent* ($\rho^2 = \rho$) and can be written $\rho = |\psi><\psi|$ for some $|\psi>$. So idempotency, uniqueness of decomposition and the possibility of writing $|\psi><\psi|$, all suffice to define the pure density matrix. Impure density matrices are said to be *mixed*.

The density matrix will be used as the concept of state. Some special features of density matrices are enlarged upon in §1.3.

1.1.2 Dichotomy of Processes

The dichotomy between Processes one and two, the first stochastic and the second deterministic, is the source of much consternation[15].

The Schrödinger equation expresses deterministic evolution where the Hamiltonian $\hat{H}(t)$ is the generator of an underlying dynamical group $U(t,t_0)$.[4] It is a group because of the property

$$U(t',t_0) = U(t',t)U(t,t_0).$$

This obeys its own Schrödinger equation,

$$i\hbar\frac{d}{dt}U(t,t_0) = \hat{H}(t)U(t,t_0)$$

the solution of which is easy for time-independent $\hat{H}$ but very difficult in general. However, the dynamical group is there and determines the state

---

[4]See [13, pp.233–240] and [26].

5

$|t>$ in terms of an initial state $|t_0>$,

$$|t>= U(t,t_0)|t_0> .$$

A simple but deep result due to Wigner, see[16], requires that the form of U be unitary([13, p.226] and [16, p.75]). Quantum evolution is required to be linear and can be represented by a unitary operator (continuity of time excludes a further anti-unitary option). The result comes from demanding that the calculated probabilities under Process One be the same under all representations of ψ. Invoking the stipulation that Process One should give the same answers to everybody, effectively demands of Process Two a linear unitary evolution.

There is some crosstalk here between apparently contradictory pictures. It can be amplified further if consideration is given to Process Two as a world picture without Process One. This can then be introduced in order to get results (observations) out of the theory. Along the way an important result due independently to Wootters and Zurek[2], and Ghirardi and Weber[3] (hereafter WZ–GW) will be reviewed. This is the Quantum No–Cloning Theorem and has consequences for the formulation of state determination problems.

A motivation of Process One is sought, assuming a quantum world described by kets evolving as per Process Two and having some correspondence with the classical world. The connection is made from consideration of the polarisation states of photons and the parametric description of the polarisation of classical e.m. waves.

At the quantum level photon polarisation can be described by a normalised linear combination of two basis states[17, p.7],

$$|\psi>= \alpha|1> +\beta|2>, \alpha\alpha^* + \beta\beta^* = 1.$$

It is well known that, after removal of the phase, states map one-one to the surface of a unit sphere in $\mathbf{R}^3$, see [18, p.71]. This is the ubiquitous *Poincaré Sphere* of quantum two-state systems, sometimes the *Bloch Sphere*. A pure quantum polarisation state is thus representable by a single point on this sphere. Single photons are observable so it is assumed single photons have states which are such points[17, p.4]. Many such identical photons, such as may be generated by a laser, are collectively described by the same single

6

point.[5]

At the classical level, an e.m. wave can be resolved into two polarisation components. The choice for this resolution is arbitrary. A pair of orthogonal states of linear polarisation, or opposite helicity states of circular polarisation are two common choices[19, p.30,31]. In a completely analagous fashion such macroscopic polarisation states are represented by points on the Poincaré sphere (invented for this purpose and transported to quantum theory). Imperfectly polarised beams of light are represented by points within the sphere.

Thus the characteristic two-stateness of luminal polarisation is represented at both levels by the same state space, a sphere.

Imagine now an apparatus that was capable of indicating the state of a single photon. As yet no connection is made between the two levels other than the sphere correspodence and the knowledge that many photons are equivalent to light as we know it. Determination of classical polarisation is easy, just rotate a polariser.[6] The way foward is clear: it is to amplify the single photon, to copy or clone it. In so doing the single point on the quantum sphere maps to a single point on the classical sphere; for which there is a measurement theory. Process Two is available to do this. The Quantum No–Cloning Theorem is concerned with limits to this procedure. Along the way Process One appears.

So assume Process Two describes evolution at the quantum level. A quantum amplifier apparatus would appear necessary. Therefore, let the initial amplifier state be $|A_0>$. The joint initial state of a single photon plus amplifier is

$$|\psi> \otimes |A_0> .$$

Process Two reigns supreme and no difficulty is encountered by postulating that a unitary evolution exists to take either of two orthogonal initial states to the appropriate multi-photon amplified states $|1;n>$ and $|2;n>$ (WZ–GW),

$$|1> \otimes |A_0> \xrightarrow{U} |1;n> \otimes |A_1>; \quad |2> \otimes |A_0> \xrightarrow{U} |2;n> \otimes |A_2> . \quad (1.1)$$

The destination amplifier states are denoted $|A_1>$ and $|A_2>$. Multi-photon states can then be mapped to the classical sphere and measurement of these

---

[5]Often a patch of points which average to an interior point of the sphere: a *mixed* density matrix[18, p.69].

[6]A generalised polariser that separates elliptical polarisations, see[1, p.73].

is understood without Process One. Problems arise when the action of this amplifier on a general input state is considered. Linearity of Process Two then demands the output

$$(\alpha|1> +\beta|2>) \otimes |A_0> \xrightarrow{U} \alpha|1; n> \otimes|A_1> +\beta|2; n> \otimes|A_2>, \qquad (1.2)$$

which is not the required state. It cannot be mapped to the classical sphere as a single point. What is needed is the amplification

$$(\alpha|1> +\beta|2>) \otimes |A_0> \xrightarrow{U} \left\{ \bigotimes_n (\alpha|1> +\beta|2>) \right\} \otimes |A_{\alpha|1>+\beta|2>}>, \qquad (1.3)$$

which does map to the classical sphere and allows direct observation of the initial seed state. This possibilty is defeated by the demand of linear evolution once the desired amplification has been specified on orthogonal basis states. It is now necessary to give an interpretation to the actual output eq.(1.2). Here the concept of *superposition* is encountered for the first time (in contradistinction to that of resolution of kets onto basis elements of a representation).[7] The actual output is a superposition of two multi-photon states. These are classically distinguishable, separated (antipodal) points of the classical sphere.

Which of the two states should be chosen? Counterfactually, it is reasonable to suppose that an alternative amplifier would have taken $\alpha|1> +\beta|2>$ to the desired classical state eq.(1.3). Observation of this with polarisers corresponding to the classical sphere-mapped images of points $|1>$ and $|2>$ would yield intensities in proportion $\alpha\alpha^*/\beta\beta^*$. This was not done, but the choice of amplifier should not affect the definition of what the photon state means. Therefore, noting the existence of *two* classical states in eq.(1.2), it is demanded that both are manifest. Each time a photon which does not conform to the amplifier prejudice is amplified either of these two classical states results with probabilities in the ratio $\alpha\alpha^*/\beta\beta^*$. This motivates Process One with the given form for the special case of two–state systems. It is also a derivation of the (WZ–GW) No–Cloning Theorem, namely that amplifiers capable of cloning arbitrary input states are impossible under the linearity constraints of quantum theory.

---

[7]Such a distinction is not necessary but serves to emphasise the sense in which superposition is often used to connote some preferred unique decomposition. This has to do with usage. Precisely, the principle refers to the normalised linear combination of two physical states as also being a physical state[17].

It is important to realise that no measurement theory is assumed here other than the knowledge of how to determine a classical polarisation state. Only a quantum world with deterministic evolution is postulated. The connection with classical physics is made by demanding that the composition of classical light as photons requires correspondence between the two isomorphic state spaces. This is a loose correspondence with a free unitary transformation[8] in-between reflecting the abstract nature of the connection.

The position can be summarised as follows:

- A consistent interpretation of Process One, meaning a representation independent definition, requires unitary evolution for the dynamical Process Two.

- Assuming just the dynamical theory, as mediated by Process Two, a stochastic assumption embodied in Process One is indicated in order that there be a correspondence between quantum states and their classical counterparts. In the case of photons the classical state is a many photon state and the existing observation theory for these (classical waves) motivates the Process One in the given form.

In no sense is this a derivation of quantum axioms. It merely serves to illustrate that the dichotomy, despite logical objections, has elements of symbiosis. In addition it is worth noting that the unknowability of single quantum states is required in order that the empirically demonstrated non-locality in EPR–type experiments[20] does not permit faster than light communication[3]. The WZ–GW result shows that within the axiomatic formulation this possibility is excluded.[9]

1.1.3 Conclusions for state determination

The principal conclusion of §1.1.2 is that quantum mechanics has a fragile logical structure which appears to dictate a law of observation that is universal to all systems $\Pr(\phi|\psi) = <\phi|\psi><\psi|\phi>$. It is the assumed universality of

---

[8]This determines an $SO(3)$ transformation; a relative rotation of the two spheres.

[9]Recently, this property of quantum systems has been used in the demonstration of a quantum communication channel that allows secure transmission of encryption keys such that any attempt to eavesdrop can be detected. This system, called $QPKD$, was developed by Charles Bennett and co–workers, it is reported in [21].

this rule which is the basis for the axiomatic formulation of measurement theory in terms of Hermitian operators. This same universality of form enables discussion of state determination without reference to any particular system or the nature of the interaction. The isomorphism of same dimension state spaces pointed out in §1.1.1 shows that the only relevant point to questions of principle is this dimensionality.

In §1.1.2 it was also shown that amplifiers are inherently prejudicial. They select a preferred basis in which to expand quantum states. The concept required here is that of an *analyser* which selects such a basis. This will be called the *measurement basis*. Exactly what kind of interaction does this is irrelevant. When the probabilties of outcomes are finally calculated the details of the interaction simply tell what this measurement basis is. Some measurement bases may not be possible, this point is ignored in order to develop limits imposed by the formalism.

Quantum states are unknowable for single systems because linearity forbids the desired amplification of general states. To determine the quantum state an ensemble of identically prepared systems is required. The same requirement applies to the measurement of expectation values. Given that the preconditions for both approaches to measurement are the same, attempted state determination is most attractive because an inference for the ensemble state enables *calculation* of the expectation value for any Hermitian operator. This point has been made before[22] in the context of state determination working from a collection of expectation value data.

The approach to state determination working with expectation values as data[23, 24, 25] is complementary to the present work. However, this method obscures the important fact that it is really only the probabilities of the various eigenvalues that carry information about the state observed. A given set of commuting Hermitian operators differ only in their eigenvalues.[10] Matching eigenvalues to eigenvectors, it is clear that all eigenvalues of different operators belonging to the same eigenvector will appear with the same probability. These probabilities are influenced by the state observed, whilst the eigenvalues reflect the choice of operator.

Therefore, eigenvalues will be discarded and it is only the basis of eigenvectors that is of interest. This is the measurement basis referred to earlier. It is realised by any non-degenerate Hermitian operator where the eigenval-

---

[10]If two matrices commute then they are simultaneously diagonalisable.

10

ues are interpreted as a way of telling which eigenvector outcome resulted. They are just labels.

For the moment note that since state determination is necessarily made over an ensemble, the possibility of mixed density matrices should be borne in mind. Having said this, pure matrices are often assumed as the only candidates for the real state. This is for special reasons and the manner for dealing with impure inputs will be indicated at the appropriate stage.

Attention now turns to defining a model for the conditions under which the state determination is to take place. Much preliminary material has been covered with a view to establishing that the problem to be discussed is well founded. This problem is interesting in its own right but may also be seen as an investigation of the limits that the universal quantum probability rule imposes upon certain kinds of information transfer. Mainstream quantum measurement theory is left from this point although there is to be no assault upon the accepted paradigm, the only difference lies in the questions asked.

1.2   The analyser-preparator couple

Consider the following model. A *preparator* ejects d–state identical subsystems in a beam. The kind of preparation involved is unimportant. Most generally it need not involve the traditional idea of a determinative measurement. It could be anything. However, often complete preparation will be assumed. In either case it should be possible to assign a density matrix to the preparator. This is what is meant by state determination. The task to be considered is that of assigning this density matrix through observation of a finite number of subsystems. Clearly the assignment will not be perfect and this is where the idea of optimality comes in. Of interest is the best way to do the assignment given a finite number, N, of available subsystems.

The beam is directed towards an *analyser*. This accepts subsystems one at a time (single quantum detection). To each it presents a *measurement basis*; this is a trial. The analyser performs traditional determinative measurements on single systems. As has been shown, details of the analyser operation are not important. It is assumed that it can be variously set to different measurement bases. Knowledge of this setting is demanded. Note that single detection is a severe practical restriction. Requesting it defines the most general problem, but it can be lifted at no cost.

11

Once a subsystem is observed the quantum outcome is noted. This outcome is one of the d eigenvectors of a measurement basis. So a datum from one trial consists of a stored analyser setting and an eigenvector result. To remove human elements this procedure could be executed by an automaton attached to the analyser. An optimal program is sought for this automaton. The program is a certain pre-chosen set of N settings which optimise the confidence with which the preparator state is known after an N–trial experiment. The sequence of settings is not important as the subsystems are indistinguishable.

Quantum results cannot be preset. They occur with probabilities calculated from the universal rule. The probabilities of a given result for all possible different preparator states are perfectly known because the analyser settings are known. There are d^N possible results for an N–trial experiment. Individual trials are independent and for imagined preparator state $|\psi><\psi|$ a result $i \in [1, d]$ can be assigned conditional probability,

$$\Pr(i|\psi) = <\psi|i><i|\psi> .$$

State determination proceeds via a natural inversion of this probability distribution called Bayes' Rule of Inference. This yields an inferred probability distribution for the preparator state fixed by the observed data. Optimality of a program is measured in terms of a kind of minimal average spread for the inverted distribution. The right measure of confidence is the Correlation Information. This concept originates from information theory and is much used in the analysis of communication systems.

Both of these concepts are given separate short chapters. This is because they may be unfamiliar to many physicists. Also, Bayes' rule is often considered trivial but there is some subtlety at work which deserves attention. Finally, all ingredients are at hand; the optimal analyser program is defined and a formal solution procedure presented. This is to be called the Optimal State Determination Problem.

Before progressing to discussion of topics from other fields it is worth enlarging upon some aspects of the space of quantum states. Probability distributions over quantum states will arise. To visualise such things a vector space model of density matrices is introduced. This will also enable a picture to be formed of the measurement bases. Alternative analyser settings appear as differently oriented geometric objects in this vector space. There are some interesting mathematical problems associated with this. The picture formed offers a natural generalisation of the Poincaré sphere for dimensionality greater than two. The material could be skipped on first reading.

1.3   Geometry of Density Matrices

Defining the density matrix mathematically as a unit trace, non-negative Hermitian matrix reveals a useful connection with the theory of vector spaces. This is well known, for collected results see [27, Chap. 2] and [28].

The idea is reviewed here with emphasis upon acquiring a picture of measurement bases and the space of states. The Poincaré sphere is derived along the way and it is possible to understand why things become complicated for higher dimension state spaces. We indicate the natural Generalised Poincaré sphere and use this to address some questions concerning special kinds of measurement bases called *mutually unbiased bases*.[11] These possess the useful property that two projectors, $|\phi_i><\phi_i|$ and $|\psi_i><\psi_i|$, drawn from different mutually unbiased measurement bases satisfy:

$$< \phi_i|\psi_j><\psi_j|\phi_i>= 1/d, \ \forall\, i,j. \tag{1.4}$$

We now explore the consequences of the joint non-negative, Hermitian and unit trace condition.

Taking Hermiticity alone first, note that any real linear combination of Hermitian matrices is also Hermitian. Therefore, they form a real vector space. For $d \times d$ matrices the dimension of this space is d^2. Notice that for two such matrices the definition

$$(A,B) \equiv \mathrm{Tr}[AB]$$

satisfies the axioms for a real inner product[31, p.437]. Standard theory of linear algebra informs us that with this inner product the $d \times d$ Hermitian matrices form a real Euclidean space. Hence they can be mapped to vector representatives in $\mathbf{R}^{d^2}$. This is nice because naive geometric intuition works in such spaces.

The inner product enables construction of bases for the vector representation. In particular it shows that orthonormal bases of vectors corresponding

---

[11] This terminology is due to Wootters[29]. They have been studied by Ivanović[30]. However, they were probably first introduced by Schwinger[27], as bases of *optimum incompatibility*.

13

to matrices can be constructed. The dimensionality is d^2 so this many basis vectors are needed. There are real angles and the usual kind of distance from coordinate geometry. A standard device defines distance in terms of the norm[12]

$$\| A \| \equiv \sqrt{(A, A)},$$

whence

$$D_{AB} = \| A - B \|.$$

So matrices have a length associated with them and there is a distance between matrices. The real Euclidean space picture shows this explicitly. There is also a real angle γ defined implicitly by writing

$$(A, B) = \| A \| \| B \| \cos \gamma.$$

If one starts to place constraints upon the Hermitian matrices then a geometry will emerge in this picture.

Start with the condition of unit trace. Call such an Hermitian matrix a pre-density matrix. Denote the unit matrix, I, by vector e and general matrix, A, by vector a. The condition of unit trace can then be written as the inner product,

$$\mathrm{Tr}[A] = \mathrm{Tr}[I \times A] \equiv (I, A) = \mathbf{e} \cdot \mathbf{a} = 1.$$

Obviously this is the condition that A lie in a plane. The dimensionality of the vector space is d^2. Such a condition confines attention to an hyperplane of dimension $d^2 - 1$. This is to be called the *trace-one plane*; denoted as P_{d^2-1}. Notice that the matrix $d^{-1}I$ has unit trace and so lies in the trace-one plane. It is clear from the defining relation for P_{d^2-1} that this happens to be the point of closest approach within this plane to the zero matrix. This can be taken as the origin, o, for the whole vector space, whilst $d^{-1}I \equiv d^{-1}\mathbf{e}$ is a useful origin, o', for points in the trace-one plane. Also, observe that

$$\| d^{-1}\mathbf{e} \| = d^{-1/2}, \text{ so that } d^{-1}\mathbf{e} = d^{-1/2}\hat{\mathbf{e}}.$$

The hyperplane, P_{d^2-1}, is by definition the set of pre-density matrices and must contain all of the density matrices. They occupy a region in this plane

---

[12]Many norms are available for matrices, see [31, p.404], this one gives a Euclidean distance.

imposed by the condition of non-negativity. A candidate pre–density matrix graduates to a full density matrix if one goes back to its matrix representation and finds no negative eigenvalues.

So imagine a painted region, an hypervolume, in P_{d^2-1} where the density matrices live.

Now let us narrow in on the density matrices. First recognise that a necessary condition for the purity of an arbitrary Hermitian pre–density matrix, ρ, is $\text{Tr}[\rho^2] = 1$. This is only sufficient for $d = 2$, a fact which is proved shortly.[13] Notice that:

$$\text{Tr}[\rho^2] = 1 \Leftrightarrow D_{\rho o} = (\rho, \rho)^{1/2} = 1.$$

This indicates that such matrices lie on the surface of a radius one hypersphere S_{d^2-1} within $\mathbf{R}^{d^2}$ and centered upon the zero matrix. Demanding both of the conditions:

$$\text{Tr}[\rho^2] = 1 \quad \text{and} \quad \text{Tr}[\rho] = 1$$

defines the intersection of P_{d^2-1} with S_{d^2-1}. This will be a second hypersphere of dimensionality one less, S_{d^2-2}, centered upon $d^{-1/2}\hat{e}$. This is contained within the trace–one plane. To verify this we simply calculate the distance of ρ from $d^{-1/2}\hat{e}$ and make use of the twin conditions given above.

$$
\begin{aligned}
D_{\rho o'} &= (\rho - d^{-1}I, \rho - d^{-1}I)^{1/2} \\
&= \left\{ \text{Tr}[\rho^2 - 2d^{-1}\rho + d^{-2}I] \right\}^{1/2} \\
&= \sqrt{1 - 1/d}.
\end{aligned}
$$

This is a constant equal to the radius of S_{d^2-2}.

The conditions used to arrive at this result are necessary for pure density matrices. Therefore they must lie on this hypersphere. Previously it was noted that general density matrices are convex combinations of pure ones. Using convexity it then follows that general density matrices are constrained to the surface and interior of S_{d^2-2}. Outside this hypersphere, and within P_{d^2-1}, there are no positive pre–density matrices and so no density matrices. It is clear that the pure states form part of the boundary to the space of states.

---

[13] Assuming sufficiency of this condition is a suprisingly common error in the literature. A simple counter–example for $d = 3$ is the matrix: $\text{Diag}(-1/3, 2/3, 2/3)$.

Of interest now is the question of whether the surface of S_{d^2-2} contains any non–density matrices. Recall the full condition for pureness is idempotency. Assume two antipodal points of S_{d^2-2} are idempotent. From this assumption a contradiction will be exhibited for all $d > 2$. To do so recognise that if ρ is one such point then its antipodal cousin is

$$d^{-1/2}\hat{\mathbf{e}} - (\rho - d^{-1/2}\hat{\mathbf{e}}) = 2d^{-1}I - \rho,$$

where we have used $d^{-1}I = d^{-1/2}\hat{\mathbf{e}}$. Squaring both matrices and using idempotency yields the two equations:

$$\rho^2 = \rho \tag{1.5}$$

$$(\frac{2}{d}I - \rho)^2 = \frac{2}{d}I - \rho. \tag{1.6}$$

The second of which reduces to

$$\frac{4}{d^2}I - \frac{4}{d}\rho + \rho^2 = \frac{2}{d}I - \rho.$$

Making use of the first equation this becomes

$$\frac{(2-d)}{d^2}I = \frac{4-2d}{2}\rho. \tag{1.7}$$

This is trivially satisfied for any idempotent ρ when $d = 2$. However, taking the trace of both sides, we obtain $(2 - d)/d = 4/d - 2$, which implies $d = 2$. It follows that for $d > 2$, idempotency at one point enforces non-positivity for the antipodal one.

This amounts to a demonstration that there are regions of the constraint sphere S_{d^2-2} which are non-positve and so remain pre–density matrices. Recalling the picture of a painted region where pre–density matrices are positive, we are saying that the sphere is not completely painted for $d > 2$, it has "holes" in it. Furthermore, a hole region is always opposite a legitimate density matrix[28]. Nevertheless, S_{d^2-2} shall be referred to as the Generalised Poincaré sphere.

A simple calculation verifies that for $d = 2$, all points on the corresponding sphere S_2 are idempotent, unit trace matrices. This amounts to a derivation of the Poincaré sphere. To make connection with the complex components of a two–state ray, $(\alpha, \beta) \in \mathbf{C}^2$, one need only form the corresponding matrix

$$(\alpha, \beta)(\alpha, \beta)^\dagger,$$

and then read off the relation between this and a general unit trace Hermitian matrix such as

$$\frac{1}{2}(1 + \hat{\mathbf{r}} \cdot \vec{\sigma}).$$

Here $\hat{\mathbf{r}} \in \mathbf{R}^3$ comprises the vector space picture of the same object within the trace–one plane.

Notice that the trace inner product is preserved by unitary transformations acting upon the matrices. It follows that any such transformation induces a rotation in the vector space $\mathbf{R}^{d^2-1}$. It cannot take us out of the trace–one plane because the trace is preserved also. This provides a simple way to see the connection between $SO(3)$ and $SU(2)$; and, because of the holes in S_{d^2-2}, why no such relation holds between $SO(d^2-1)$ and $SU(d)$ for $d > 2$.

The general picture that results is shown in figure(1.1). We now turn to consideration of how complete bases appear in this picture. To pursue this note that orthogonality of two projectors $P = |\psi><\psi|$ and $Q = |\phi><\phi|$ is equivalent to the condition:

$$\mathrm{Tr}[PQ] = \mathbf{p} \cdot \mathbf{q} = 0.$$

This shows that the vector representatives of orthogonal projectors are also orthogonal in the whole space. If one takes just the components of these in the trace–one plane then we find that

$$\mathrm{Tr}[(P - d^{-1}I)(Q - d^{-1}I)] = (\mathbf{p} - d^{-1/2}\hat{\mathbf{e}}) \cdot (\mathbf{q} - d^{-1/2}\hat{\mathbf{e}}) = -1/d.$$

Since the d elements of a basis are mutually orthogonal it follows that their projections in the trace–one plane make the same angle with each other. Recall the radius of the sphere is $(1 - 1/d)^{1/2}$ so that this angle is

$$\cos\gamma = -1/d \times (1 - 1/d)^{-1} = -1/(d-1).$$

Furthermore, projectors must lie on the Generalised Poincaré sphere. The mutual equidistant property of an orthonormal set thus defines a d-vertex, dimension $d-1$ generalised tetrahedron. This lies within the trace–one plane and its vertices touch the Generalised Poincaré sphere.

Such an object will be called a d-vertex equi–pyramid and one corresponds to each measurement basis. For two states the pyramid degenerates

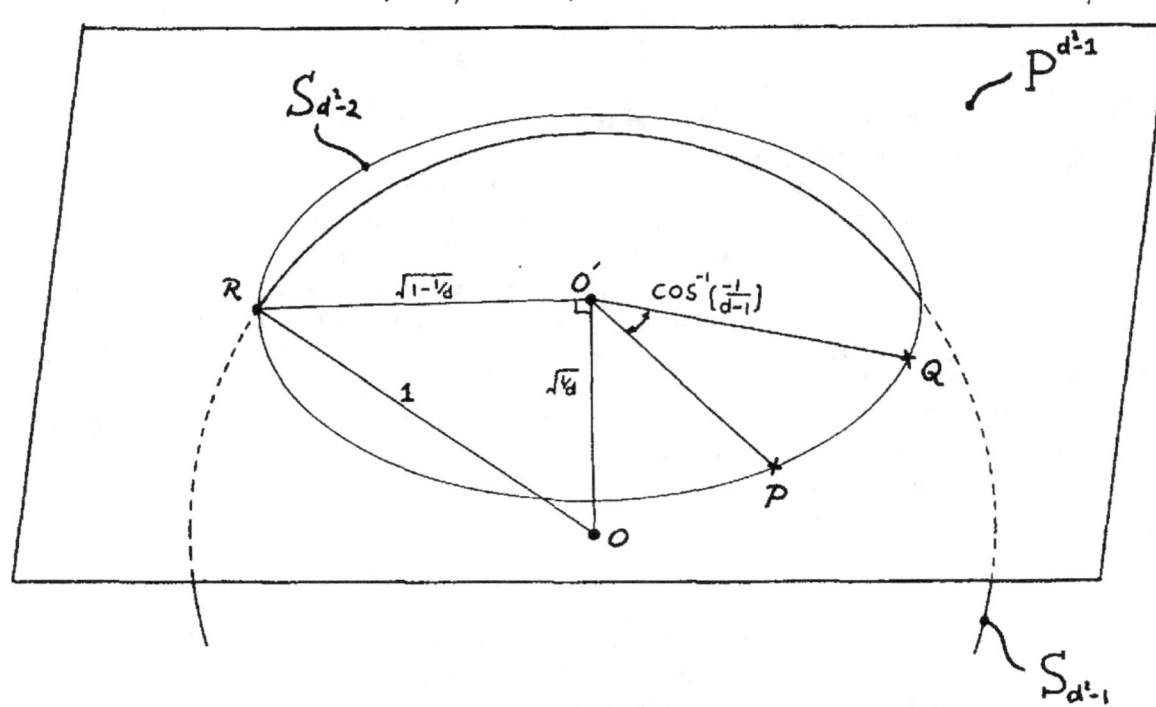

Figure 1.1: Generalised Poincaré sphere lying in the d^2–dimensional real vector space of $d \times d$ Hermitian matrices. The trace–one plane, P_{d^2-1} contains this sphere. Orthogonality of two projectors in the whole space becomes an angular seperation of $\cos^{-1}(-1/(d-1))$ for the components within this plane. Mutually unbiased projectors have orthogonal components in the plane while a complete set of orthogonal projectors defines a d–vertex pyramid that touches only positive portions of the Generalised Poincaré sphere.

18

to a line segment ($\cos \gamma = -1$). Then we recover the familiar picture of basis element pairs mapping to antipodal points of the Poincaré sphere.

Different choices of equi–pyramid correspond to different measurement bases. However, not all pyramids are measurement bases. This is because for $d > 2$, some vertices may lie on non–positive regions of S_{d^2-2}. This has consequences for the properties of mutually unbiased bases, as defined in eq.(1.4).

For two projectors, P and Q which are mutually unbiased, $\text{Tr}[PQ] = 1/d$, a repeat of the above argument shows that

$$\text{Tr}[(P - d^{-1}I)(Q - d^{-1}I)] = (\mathbf{p} - d^{-1/2}\hat{\mathbf{e}}) \cdot (\mathbf{q} - d^{-1/2}\hat{\mathbf{e}}) = 0.$$

Geometrically, mutually unbiased projectors have orthogonal trace–one plane components. It follows that mutually unbiased bases must have pyramids that lie in orthogonal subspaces of the plane P_{d^2-1}.

If we now recall that the dimensionality of this plane is $d^2 - 1$, and recognise that each pyramid is of dimension $d - 1$, then it is clear that at most $d + 1$ orthogonal subspaces are available to form no more than this number of mutually unbiased bases.[14] For two states one expects at most three such bases and these are given by the eigenvectors of the Pauli matrices.

An explicit construction of $d + 1$ mutually unbiased bases has been given for the case of prime dimensionality by Ivanovič[30].[15] For composite d, mutually unbiased bases may still be formed. However, there are fewer than $d + 1$. This suprising importance of prime numbers may be traced to certain properties of Gauss sums[27, 32]. Although the partition of P_{d^2-1} into $d + 1$ orthogonal $d - 1$ dimensional subspaces is always possible, the nature of the positive region on the Generalised Poincaré sphere does not allow the insertion of the maximal number of d–vertex pyramids, that strike everywhere positive regions, unless d is a prime number.

When we come to do calculations for higher spaces this feature will cause us to concentrate attention upon prime dimensional spaces for certain questions. The picture arrived at is used to aid visualisation of the nature of biased measurement bases as differently oriented objects in a vector space. It will also help us to understand why there is an essential difference between calculations for two–states and those for $d > 2$.

---

[14]This answers a question raised by Wootters[29].

[15]He observed that they have the property of admitting an orthogonal decomposition of state space along the lines we have indicated above.

Chapter 2

Bayesian Inference

2.1   A simple measurement model

The following simple and highly general model of measurement will be considered. Given are a system S and apparatus A. The system variables are denoted x_S lying in some abstract state space X_S, whilst those of the apparatus are denoted x_A lying in state space X_A. In general these spaces may be quite different. For instance, one may be continuous whilst the other is discrete. Possible restrictions on the nature of these spaces are not of interest because the model works in the cases important to this study.

It is to be in the nature of an apparatus that there is perfect knowledge of its reading. A reading is just a specification of some sharp value for x_A. Measurement of S by A involves introducing a *correlation* between the two such that system variables and apparatus variables are related by a conditional probability distribution,

$$p(x_A|x_S). \tag{2.1}$$

How this correlation is brought about is of no concern. It will be as the result of some interaction. It is not necessary to know this in order to develop a theory of measurement. The interaction lies buried in the conditional probability. A theory is good enough to develop a notion of measurement if the conditional probability is available.

Therefore, for the purpose of measurement all knowledge of the interaction reduces to the essential statement:

If the system state is x_S then the apparatus state is, or will become, x_A with probability $p(x_A|x_S)$.

Notice that the requirement of perfect knowledge of the apparatus variables presents no serious constraint. Any uncertainty about the reading may be incorporated into the conditional probability distribution. Very general apparatus can be dealt with in this manner. The word apparatus remains undefined but its scope may extend right up to the level of the human mind. It is something that can be read according to the above definition.

Measurement is not normally discussed in this way. In physics an experimenter often attempts to construct an apparatus whose apparatus state space is *isomorphic* to the state space of the system under study. Generally the real numbers are involved so specialise to the simple case $X_A, X_S \cong \mathbf{R}$. Within the present scheme an ideal apparatus then has a correlation function of delta function character,

$$p(x_A|x_S) = \delta(x_A - x_S).$$

In such circumstances the system state can be read off as the apparatus state. Such a simplification leads to a merging of the two spaces—this is to be avoided to understand measurement in full generality.

Noise is ever present in nature and imperfect correlation is the norm. If similar state spaces are retained then a very wide class of measurements can be modelled by correlations of Gaussian form. A single parameter takes care of the noise. That parameter is the standard error of a single measurement, denoted σ. The measurement correlation is then

$$p(x_A|x_S) = \frac{1}{\sqrt{2\pi}\sigma} \exp\left\{ -\frac{1}{2}\left(\frac{x_A - x_S}{\sigma}\right)^2 \right\}. \qquad (2.2)$$

In a single observation the result of the measurement is

$$x_S \backsimeq x_A \pm \sigma.$$

Often there is a possibility of repeated observation of the system with the same apparatus. Upon repeated observation the result becomes

$$x_S \backsimeq \bar{x}_A \pm \frac{\sigma}{\sqrt{N}}.$$

Here N trials are made and the system is assumed to remain relatively undisturbed, or there are identical copies available. Of course the uncertainty is reduced by new data.

Practical examples of such measurements include:

- Measurement of intervals with a ruler where σ describes a very complicated interaction between human and environment.

- Measurement of voltages with a multimeter where σ essentially describes the electric circuit—meter needle interaction.

Suitably abstract system and apparatus state spaces can be constructed in both cases. However, the examples are too familiar for this to be anything more than an exercise in perversity. The point is that the indicated model describes such classical measurements. Normally it is not needed because the inversions from observed x_A to inferred x_S are trivial or at least highly familiar.

Having defined an abstract measurement correlation in probabilistic terms there is a suggestive possibility for general inversion. Given perfectly known x_A, find a probability distribution for x_S. This is easily expressed as a reversal of the conditional probabilty.

Given $p(x_A|x_S)$, obtain $p(x_S|x_A)$.

The theory of Bayesian inference is concerned with this problem. Notice that the inverted probability distribution is usually rather more than is required. It can be reduced to an *estimator* of the system state. In the above example of Gaussian correlations the mean has been used and the uncertainty implicit in the distribution over the system state is captured by σ.

Bayesian inference will now be described with special emphasis on an aspect that often engenders confusion. This is the concept of a prior distribution.

2.2   Bayes' Rule of Inference

Bayesian inference provides a general method for inverting conditional probability distributions. It is indicated by Bayes' theorem[33, p.74]. This defines conditional probabilities in terms of a joint probability, say $p(A, B)$, and its marginals, $p(A) = \sum_B p(A, B)$ and $p(B) = \sum_A p(A, B)$.[1] A compact form for the case of two random variables follows:

$$p(B|A)p(A) = p(A, B) = p(A|B)p(B). \qquad (2.3)$$

Normally, $p(A, B)$ is given and so this defines the two possible conditional probability distributions. However, in the case where one conditional is known it makes sense to think about eq.(2.3) as determining the other conditional. Suppose $p(A|B)$ is known. Then specifying $p_0(B)$, *by whatever means*, defines $p(A, B)$ and, in a self-consistent way, determines $p(A)$ and $p(B|A)$. Making use of the above properties alone these are:

$$p(A) = \sum_B p(A|B)p_0(B) \qquad (2.4)$$

$$p(B|A) = \frac{p(A|B)p_0(B)}{\sum_B p(A|B)p_0(B)}. \qquad (2.5)$$

This procedure is called Bayes' Rule of Inference. All distributions so generated are correctly normalised in virtue of Bayes' theorem. Under the given conditions the inversion is non-unique, being influenced by the choice of $p_0(B)$. Any confusion that exists from the application of Bayesian inference stems from this non-uniqueness. In view of its special role $p_0(B)$ is called a *prior* distribution (with subscript zero to emphasise this). Correspondingly, $p(B|A)$ is called a *posterior* distribution for B.

Trivial substitutions: $x_A \leftrightarrow A$ and $x_S \leftrightarrow B$, establish connection with the inversion problem of the measurement model described above. Applying eq.(2.5) turns the measurement correlation $p(x_A|x_S)$ into an inferred distribution for x_S parametrised by the observed data x_A, namely: $p(x_S|x_A)$.

The simple measurement model indicates that it is sensible to demand that $p(A|B)$ be known. An *interpretation* can then be given to $p_0(B)$ as summarising the prior knowledge about what value B takes. This knowledge might be acquired by any means. An example might be that $B \in [0, 1] \subset \mathbf{R}$ as a matter of necessity. Restrictions of this kind are the most important, even though they do not select a unique functional form. Often mention is made of the need for principles that aim to do this. Rather than debate that point it is more enlightening to ask to what extent the functional form really matters.

---

[1]Summation stands for integration as well; a mixture of discrete and continuous variables is allowed.

2.3 Properties of the Prior Distribution

A short argument will be given to show that the functional form of the prior is only important for small numbers of observations. For large numbers of observations only the a priori restrictions on allowed values for B are relevant. The idea for this originates from a discussion in a book by von Mises[34]. The argument given here is complementary to that of von Mises. A physical approach is adopted by appealing to scale invariance present in the Bayesian inversion formula. No attempt is made at rigour and in mind are correlations defined in terms of functions on real numbers although this need not be necessary.

To begin with note that eq.(2.5) represents the best inference that can be made in a single observation consistent with $p_0(B)$. Unless the measurement correlation is strongly peaked it will matter what the functional form of the prior is. For poor correlations individual observations do not contribute very much. It is then no surprise that results are highly dependent upon initial knowledge. Good information from small numbers of observations requires a sharp measurement correlation.

With a poor measurement correlation the way to improve knowledge is to repeat observations. Then either one needs non-disturbing measurement or an ensemble of identical systems. Either way Bayes' rule provides an intuitively attractive picture of such chained inference. Consider then a fixed correlation $p(A|B)$, select a prior $p_0(B)$, and imagine N independent trials to determine what B is the most likely cause of observed data $\{A_i\}_{i=1}^N$. Bayes' rule chains naturally with inferred distributions from the ith trial forming the prior for the $i + 1^{th}$ trial. After N trials the inference for B is:

$$p(B|\{A_i\}_{i=1}^N) = \frac{\left(\prod_{i=1}^N p(A_i|B)\right) p_0(B)}{\sum_B \left(\prod_{i=1}^N p(A_i|B)\right) p_0(B)}, \qquad (2.6)$$

which can be thought of as Bayes' rule applied to a new N–trial correlation

$$p(\{A_i\}_{i=1}^N|B) \equiv \prod_{i=1}^N p(A_i|B), \qquad (2.7)$$

with the original prior $p_0(B)$ retained.

Now two properties of eq.(2.6) are of interest. Rescaling the prior does not change the posterior distribution. Also wherever the prior is zero the posterior is zero.[2] Taking the second property first, any finite measure region with $p_0(B)$ equal to zero is a priori excluded as an inference. The rescaling property proves useful in the large N limit.

For large N it is reasonable to assume that for fixed outcome $\{A_i\}_{i=1}^N$ the product in eq.(2.7) becomes strongly peaked about some point $B_{\max}$. Except in cases of special symmetry the product guarantees that one value of B will dominate.[3] Referring now to eq.(2.6) think of $p(\{A_i\}_{i=1}^N | B)$ as a peaky function that samples $p_0(B)$. In particular it samples values around $B_{\max}$. For this value of the outcome $\{A_i\}_{i=1}^N$, the peakedness of such sampling enables the rest of the prior to be discarded. Then rescaling says that for $p_0(B)$ locally flat compared to such a sampling the actual value does not matter. It only affects the tails of the inferred distribution. The same is true for all samplings $B_{\max}$ where each $\{A_i\}_{i=1}^N$ generates a different one. So although $p_0(B)$ may be a bumpy function, the rescaling property ensures that for large enough N the N–trial correlation is sufficiently peaked to sample only its curvature and not its value. The actual value only matters when more than one peak is present. Notice that this argument could be rephrased slightly to show that for large N, further trials do not add much confidence to the inferred distribution.

In conclusion:

- The functional form of the prior is only important for small numbers of trials when not much is known about B.

- Regions of B–space where the prior is zero are *always* significant and exclude certain inferences a priori.

Where large numbers of trials are possible then the functional form of the prior becomes progressively less important. In such circumstances it makes sense to choose a prior which is constant, since this allows the observations to express the true B with least interference. This choice of uniform prior is often differently motivated as encoding complete ignorance about the value of B. Such a choice is often elevated to a principle: *"Laplace's principle of insufficient reason"*. It could be claimed that the above argument shows this principle to embody that choice consistent with the asymptotic behaviour of Bayes' rule.

---

[2]In some circumstances this restriction may be relaxed slightly using L'Hospital's rule.

[3]Assuming appropriate conditions upon the fundamental correlation $p(A|B)$.

Either way, it should be clear that there are very good reasons for choosing a uniform prior. An added bonus is that inversion becomes particularly simple for uniform prior. For example, specialising to a real valued x_S yields:

$$p(x_S|x_A) = \frac{p(x_A|x_S)}{\int p(x_A|x_S)dx_S}. \tag{2.8}$$

This is just the measurement correlation itself with appropriate normalisation.

2.4 Examples

The simple measurement model has been formulated to include both classical measurements and the special problem of quantum state determination using an analyser-preparator couple. Introduction of the quantum mechanical result will be delayed until after the next chapter.

At this stage it is worth fixing ideas with two simple examples. First, a derivation is given of the rule of diminishing error for Gaussian correlations. Second, consideration is given to the problem of measuring the *probabilities* for an unfair coin. This is a very useful prototype for the quantum state determination problem. It is also the problem Bayes considered in the process of developing his theory[34].

2.4.1 Gaussian correlation

It is instructive to derive the standard results of error analysis in a Bayesian context. Pick a Gaussian correlation with standard error σ as in eq.(2.2) but relabel $x_A \leftrightarrow x$ and $x_S \leftrightarrow x'$. So the correlation is,

$$p(x|x') = \frac{1}{\sqrt{2\pi}\sigma} \cdot \exp\left\{-\frac{1}{2}\left(\frac{x-x'}{\sigma}\right)^2\right\}. \tag{2.9}$$

A uniform prior can be included by choosing integration with uniform measure on $(-\infty, +\infty)$. For a single observation application of the form eq.(2.8) yields the same expression as above. So $x' = x \pm \sigma$ after one trial.

After N independent trials, denote the observed data by $\mathbf{x} = \{x_i\}_{i=1}^N$. Then application of eq.(2.7) enables interpretation of the repeated trials as defining a new correlation. This is

$$p(\mathbf{x}|x') \equiv \prod_{i=1}^{N} p(x_i|x').$$

(2.10)

Inversion proceeds as before on this new correlation with the original uniform prior. It is only necessary to calculate the appropriate normalisation. Substituting for the explicit form of the correlation this is

$$\int_{-\infty}^{+\infty} \exp\left\{-\frac{1}{2}\sum_{i=1}^{N}\left(\frac{x_i - x'}{\sigma}\right)^2\right\} dx'.$$

(2.11)

Doing this elementary integral followed by some minor manipulation yields

$$p(x'|\mathbf{x}) = \frac{1}{\sqrt{2\pi}\sigma'} \cdot \exp\left\{-\frac{1}{2}\left(\frac{x' - \overline{x}}{\sigma'}\right)^2\right\}.$$

(2.12)

where

$$\sigma' = \frac{\sigma}{\sqrt{N}}, \overline{x} = \frac{1}{N}\sum_{i=1}^{N}x_i.$$

Thus Bayesian inference provides a simple derivation of the familiar result

$$x' = \overline{x} \pm \frac{\sigma}{\sqrt{N}}.$$

Notice that the N trial outcome $\mathbf{x} = \{x_i\}_{i=1}^{N}$ reduces to the simple estimator $\overline{x}$, the familiar arithmetic mean of the data.

2.4.2 Coin probabilities

The original problem studied by Bayes involved measuring the bias of an unfair coin. It is useful to explore this as the quantum state determination problem is very similar. The system state space is rather abstract, being the set of all possible probabilities for heads. To simplify calculation it is assumed that all values in $[0,1]$ are possible and equally likely. The apparatus consists of the action of tossing the coin and noting the outcome Head or Tails. Therefore, take

$$X_S = [0,1] \quad \text{and} \quad x_S = q \in [0,1],$$
$$X_A = \{H,T\} \quad \text{and} \quad x_A = H \text{ or } T.$$

27

The correlation has the simple form

$$\begin{aligned} \Pr(H|q) &= q, \\ \Pr(T|q) &= 1-q. \end{aligned}$$

and a uniform prior distribution $p_0(q) = 1$ is assumed. Under such conditions eq.(2.8) applies.

Of course a single throw cannot through a binary outcome express the continuum of possible values for q. However, repeated trials make this possible and these are available with a coin.

For repeated trials only the number of heads observed is important. Suppose this number of heads is n then define a new variable $Q = n/N$. This new outcome variable takes values on a lattice of points between zero and one, of spacing $1/N$. As $N \to \infty$ this lattice becomes ever more dense, simulating the continuous interval $[0, 1]$.

In terms of this new variable the N–trial correlation is

$$p(Q|q) = \begin{pmatrix} N \\ NQ \end{pmatrix} q^{NQ}(1-q)^{N(1-Q)}.$$

Inversion of this, noting uniformity of the prior, gives

$$p(q|Q) = (N+1) \times \begin{pmatrix} N \\ NQ \end{pmatrix} q^{NQ}(1-q)^{N(1-Q)}. \tag{2.13}$$

Here use has been made of the fact that

$$\int_0^1 q^{NQ}(1-q)^{N(1-Q)}dq = \left[(N+1) \times \begin{pmatrix} N \\ NQ \end{pmatrix} \right]^{-1}$$

for integer values of NQ. This comes from identifying the above integral as a Beta function[35, p.950]. Further properties of this integral enable calculation of two useful parameters. The mean for inferred q in terms of data $Q = n/N$ is

$$\bar{q} = \frac{NQ+1}{N+2} = \frac{n+1}{N+2}$$

28

where n is the number of heads. The variance is

$$\sigma^2 = \frac{(NQ+1)(N(1-Q)+1)}{(N+3)(N+2)^2},$$

which simplifies to

$$\sigma^2 \sim \frac{Q(1-Q)}{N} + O(1/N^2)\ldots$$

The results can be summarised in the intuitively obvious conclusion that the probability is measured by taking the limit of the frequency of heads. Of value is the additional information that the variance, measuring confidence in the result, depends on both N and the actual coin probability. In particular for Q close to unity or zero the variance is least. Heavily biased coins are easier to distinguish from their neighbours. This point has been discussed before by Wootters[1].

Increased confidence for certain regions of the system state space is a general feature of these inference problems. When the quantum problem is considered it turns out that there is sufficient freedom in the single trial correlation so as to enable the N–trial correlation to be chosen to minimise, or even remove this effect. This is done by specifying different measurement bases. To measure the optimality of various possible N–trial correlations we shall use a quantity known as the Correlation Information.

Chapter 3

Correlation Information, Communication and Measurement

3.1 Information theory

The subject of Information Theory has its origins in the theory of communication systems and is finding increasingly varied application in many areas of physics[5, 6, 7, 36, 37, 38, 39, 40]. There are many excellent books on the subject, see [41, 42, 43, 44], so we shall be content with developing the theory in terms of the properties of a fundamental quantity called the Information Entropy. This is a measure of the uncertainty represented by a probability distribution. Often the uncertainty is genuinely related to questions of incomplete knowledge, although the quantity itself is merely a measure of spread. In this respect it is similar to variance but superior for several reasons.

The starting point for all treatments is a discrete probability distribution on N outcomes, $p_i \geq 0$, $\sum_{i=1}^{N} p_i = 1$. To this is associated the positive number:

$$H_N(p_1, \ldots, p_N) = -\sum_{i=1}^{N} p_i \log p_i. \tag{3.1}$$

This is the familiar entropy of thermodynamics, albeit with a different constant and a slightly different interpretation. For this reason it is called the Information Entropy. It was first introduced by Shannon[41], who wished to

30

quantify the information passed by communication systems.

The simplest possible communication system consists of a transmitter signalling a receiver sequentially from a possible set of N discrete symbols. The necessary physical connection between the two is called a channel. Shannon reasoned that a mathematical measure of the information content of a string of such symbols should have nothing to do with the semantic content of them as a message. In particular the nature of the symbols themselves is unimportant. Communication is effected by simply recognising one of N possible alternatives when a symbol is received. From this, semantic content is recovered by knowing a code, but the actual information transmitted depends only upon the frequency with which symbols are transmitted.

Therefore Shannon proposed to measure such information in terms of a quantity associated with discrete probability distributions. The information entropy has the useful property of convexity (inherited from its $p \log p$ form) which results in the following inequality,

$$0 \leq H_N(p_1, \ldots, p_N) \leq \log N. \tag{3.2}$$

The extrema are attained by particular distributions.

The maximum is attained by a uniform distribution, $p_i = 1/N$. If such a probability distribution is thought of as describing ignorance about which of N options is really so, then this represents maximum uncertainty. Alternatively, for the communication model, a code that ensures such a probability for transmitted symbols seems intuitively the most efficient in virtue of the equal employ of all symbols (which obviously have the same coding value). Both pictures are united if one imagines the distribution over transmitted symbols as the receiver's prior distribution for what symbol comes next. On reception this prior collapses to certainty and so the maximum possible uncertainty has been removed.

The minimum is attained by a distribution concentrated on a single outcome, $p_i = 1$ for some i. Clearly, this corresponds to zero uncertainty. Also, a transmitter that sends one symbol constantly is not being very informative. Perfect reception involves a transition from some prior distribution between these extremes to a posterior distribution of certainty. In the previous paragraph the situation described results in the change from a priori maximal uncertainty to a posteriori minimal uncertainty. This turns out to be the optimal channel for the simple situation discussed. Therefore, with perfect reception the information transmitted per symbol is measured by the Information Entropy of the transmitter source probability p_i.

Shannon made a guess at the Information Entropy but the great usefulness of his work stems from the fact that he then motivated this choice upon axiomatic grounds. The axioms of information theory, in their various forms, seek to give reasonable a priori conditions on a family of functions $H_N(p_1, \ldots, p_N)$ (N considered variable) such as to constrain the choice.[1] It is then a theorem of information theory that the choice is uniquely the Information Entropy; hence the new name.

3.2   Communication and Correlation Information

Results of the last chapter on Bayesian inference enable immediate transition to a model of a communication channel with noise[41, p.5]. The resulting picture will also connect the notions of communication and measurement. It turns out that the efficacy of both is measured by the Correlation Information, derived in a natural way from the Information Entropy.

Noise in a discrete channel[2] communication system can be modelled by introducing a probabilistic receiver condition. Let the source probability of transmitted symbols T_i be $p_0(T_i) = p_i$. It plays the role of a Bayesian prior in what follows but notice that it is a genuine known probability distribution set by the way messages are coded into symbols. Each T_i has a corresponding received symbol R_i. The nature of channel noise is expressed by choosing some conditional $p(R|T)$. The situation is indicated schematically in figure(3.1).

The prior uncertainty about T at the receiver is

$$H_{\text{prior}} = - \sum_T p_0(T) \log p_0(T). \tag{3.3}$$

Bayesian inference at the receiver gives

$$p(T|R) = \frac{p(R|T)p_0(T)}{\sum_T p(R|T)p_0(T)},$$

---

[1]This is not empty axiomatics but a true codification of the intuitive notion of information, in particular an additive property, see [41, 42].

[2]This means communication with a finite inventory of signal symbols.

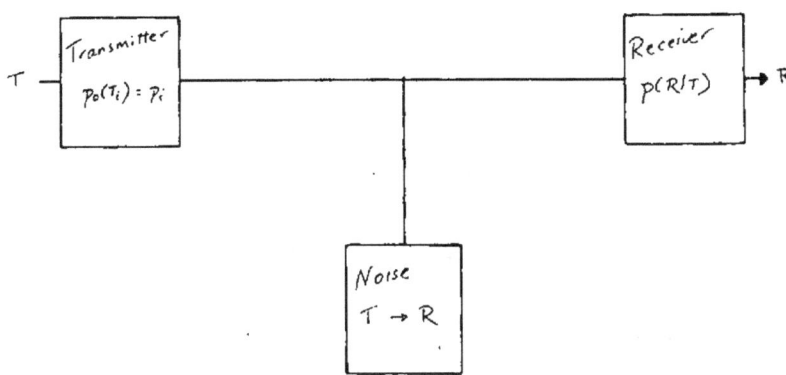

Figure 3.1: Schematic for a discrete communication channel. Symbols T from the transmitter are decoded, with some error due to noise, into symbols R with probability $p(R|T)$. The code is such that transmitter symbols are sent with frequency $p_0(T)$.

33

where symbols R occur with probability

$$p(R) = \sum_T p(R|T)p_0(T).$$

To this is assigned the posterior uncertainty,

$$H_{\text{posterior}}(R) = -\sum_T p(T|R) \log p(T|R). \tag{3.4}$$

Notice that this depends upon the symbol received. The posterior uncertainty about T averaged over all possible received symbols is

$$H_{\text{posterior}} = -\sum_R p(R) H_{\text{posterior}}(R). \tag{3.5}$$

Substitution from the previous equation, noting that $p(T,R) \equiv p(T|R)p(R)$, results in the compact form

$$H_{\text{posterior}} = -\sum_{T,R} p(T,R) \log p(T|R). \tag{3.6}$$

Subtracting the prior uncertainty from the average posterior uncertainty measures the change in uncertainty represented by the information transmitted down the channel. It will be *loss* of uncertainty representing a *gain* of information. Therefore,

$$\overline{\Delta I} = -\overline{\Delta U} = H_{\text{prior}} - H_{\text{posterior}}.$$

Substitution from the above expressions shows that the average information gained per symbol received is

$$\overline{\Delta I} = -\sum_T p(T) \log p(T) + \sum_{T,R} p(T,R) \log p(T|R). \tag{3.7}$$

But $p(T) = \sum_R p(T,R)$ and so after some manipulation this reduces to the more useful form

$$\overline{\Delta I} = +\sum_{T,R} p(T,R) \log \left[\frac{p(T,R)}{p(T)p(R)} \right]. \tag{3.8}$$

The above argument represents a motivation for elevating this equation to the status of a definition for a new derived quantity called the Correlation

34

Information, denoted $\{T, R\}$. In the context of communication this measures the information carrying capacity of a channel in units of *nats* (base–two logarithms for the more familiar *bits*). It can be shown that this is always non-negative.

$$\{T, R\} = + \sum_{T, R} p(T, R) \log \left[\frac{p(T, R)}{p(T)p(R)} \right] \geq 0.$$

This expresses the intuitive fact that information is non-decreasing at the receiver. Something or nothing is learned on receipt of each symbol but cannot be un-learned, so to speak. Further properties of the Correlation Information will be developed in a later section. For the moment we concentrate upon an obvious connection between communication channels and the model of measurement developed in the last chapter.

3.3 Communication and Measurement

Any problem of Bayesian inference will have associated with it a Correlation Information. This is because specifying a conditional $p(A|B)$ and prior $p_0(B)$ defines a joint distribution $p(A, B)$ and therefore a Correlation Information through use of eq.(3.8). This Correlation Information measures the information gained about B from knowing A or conversely, in virtue of the manifest symmetry of eq.(3.8), what is learned about A from knowing B.

In the communication model given above, $p(R|T)$ could be considered a physical property of a given channel whereas $p_0(T)$ is influenced by the way a message is coded. The quantity $\{T, R\}$ measures the optimality of the system and one might seek a coding that results in a $p_0(T)$ which maximises this.

Similarly there is a Correlation Information associated with the measurement problem as defined in the last chapter. This suggests an interpretation of measurement as a certain form of communication between system and apparatus, figure(3.2). However, as far as optimality is concerned the emphasis is slightly different.

The measurement correlation $p(x_A|x_S)$ is influenced by the choice of apparatus which will be considered as open to influence. However, the prior $p_0(x_S)$ stands for initial information about the system state. This should

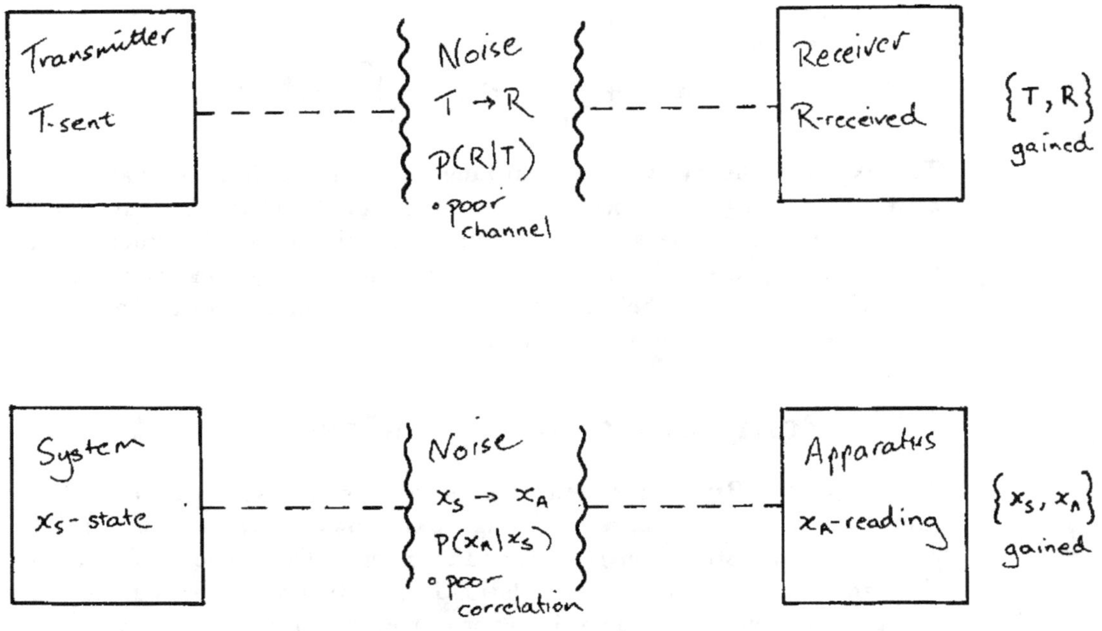

Figure 3.2: Schematics for communication and measurement. We wish to emphasise that the efficiency of both is measured by the Correlation Information.

be considered fixed if the performance of different apparatus is to be compared. A logical choice is the uniform prior since this averages performance evenly over all possible system inputs. If selective comparison of apparatus over some subset of likely inputs is required then a prior peaked on that set should be chosen. In any case optimality is still measured by the Correlation Information $\{x_A, x_S\}$. The optimal measurement problem involves varying $p(x_A|x_S)$ in order to maximise this.

Thus Bayesian inference supports questions of optimality for both communication and measurement through a common quantity, the Correlation Information. Now attention concentrates upon some properties of this.

3.4 Properties of the Correlation Information

Many different notations are used in the literature. In this work the notation of Everett is adopted. Chapter two of his thesis[45] on relative states in quantum mechanics (many-worlds interpretation) provides a valuable reference on properties of the Correlation Information in a physics environment.

First note that the manifest symmetry of eq.(3.8) in the two variables R and T suggests a natural generalisation of the Correlation Information to *any* joint probability distribution on an arbitrary number of variables. Associated to $p(X, Y, \ldots, Z)$ is the Correlation Information

$$\{X, Y, \ldots, Z\} = \sum_{X,Y,\ldots,Z} p(X, Y, \ldots, Z) \log \left[\frac{p(X, Y, \ldots, Z)}{p(X)(Y) \cdots p(Z)} \right], \quad (3.9)$$

which satisfies

$$\{X, Y, \ldots, Z\} \geq 0, \quad (3.10)$$

where

$$\{X, Y, \ldots, Z\} = 0 \iff p(X, Y, \ldots, Z) = p(X)p(Y) \cdots p(Z). \quad (3.11)$$

Thus the Correlation Information is non-negative and zero if and only if the variables are statistically independent. Aside from any particular association with physical problems it therefore represents a fundamental measure of correlation between variables. In this respect it is superior to other statistical measures such as the covariance.

37

Notice that eq.(3.9) can be rewritten in terms of the Information[3] of the total distribution and its marginals. Returning to the case of two variables (with obvious generalisation) define Informations:

$$
\begin{aligned}
I_{A|B} &= \sum_{AB} p(A,B) \log p(A|B), \\
I_B &= \sum_{AB} p(A,B) \log p(B), \\
I_{AB} &= \sum_{AB} p(A,B) \log p(A,B), \\
I_{B|A} &= \sum_{AB} p(A,B) \log p(B|A), \\
I_A &= \sum_{AB} p(A,B) \log p(A).
\end{aligned}
\tag{3.12}
$$

Then it is simple to verify that,

$$
I_{A|B} + I_B = I_{AB} = I_{B|A} + I_A.
\tag{3.13}
$$

Working from the corresponding version of eq.(3.9) it is then clear that the Correlation Information can be expressed in three different ways:

$$
\{A,B\} = I_{AB} - I_A - I_B,
\tag{3.14}
$$

$$
\{A,B\} = I_{A|B} - I_A,
\tag{3.15}
$$

$$
\{A,B\} = I_{B|A} - I_B.
\tag{3.16}
$$

The first expresses the symmetric role played by both variables whilst either of the other two may prove useful for explicit calculation.

The last question of interest to us concerns generalisation of the Information Entropy to continuous probability densities. This possibility was glossed over in the last section because there are problems of uniqueness. To see this, consider the obvious extension of $H_N(p_1, \ldots, p_N)$ to a continuous density $\rho(x)$,

$$
H_\rho = - \int \rho(x) \log \rho(x) dx
\tag{3.17}
$$

where $\int \rho(x) dx = 1$. Now there is an interesting difficulty associated with interpreting this as measuring the uncertainty about x. Any differentiable

---

[3]It is a matter of convenience to take Information as minus the Information Entropy, although there is considerable divergence in nomenclature.

one-one map of the variable x, of which rescalings are but one example, defines a new density with different Information Entropy. But such an invertible map could not be considered as adding anything to our knowledge.

For example, take $y = f(x)$ where f has inverse g such that $x = g(y)$ then identify $dx = d[g(y)] = g'(y)dy$. Under this map a new density is induced $\tau(y)$, where $\tau(y) \equiv \rho[g(y)] \cdot g'(y)$ and $\int \tau(y)dy = 1$. Taking the natural definition

$$H_\tau = -\int \tau(y) \log \tau(y) dy,$$

a simple change of integration variables shows that H_ρ and H_τ are related by,

$$H_\tau = H_\rho - \int \tau(y) \log g'(y) dy. \qquad (3.18)$$

In general a different information will result despite the fact that the two densities are essentially different representations of the same data. The ability to vary the information by such a trick shows that it cannot measure an absolute property of the equivalent family of densities. The same is of course true of statistical measures like the variance. This problem is not encountered for the discrete distribution because the invertible maps are then just permutations of the labels and H_N is trivially invariant.

This would appear to be a difficulty until the Correlation Information of a joint probability density is formed. Then it is found that this so-called measure dependence of the Information Entropy cancels. Importantly, the Correlation Information is therefore an *absolute* property of joint distributions and densities. The definition given in eq.(3.9) has the obvious extension to continuous densities wherein the right-hand side becomes,

$$\int p(X, Y, \ldots, Z) \log \left[\frac{p(X, Y, \ldots, Z)}{p(X)(Y) \cdots p(Z)} \right] dX \, dY \cdots dZ. \qquad (3.19)$$

Clearly, mixtures of continuous and discrete variables are also allowed.

Everett has given an excellent discussion and proof of the invariance of the Correlation Information under invertible maps of the variables. Here it is enough to point out that it comes from cancellation of the measure dependent factors between numerator and denominator of the log argument. Hence it is most easily proved by working with the above form symmetric in all variables. Measure theory is necessary for a complete proof although making a simple invertible, differentiable change of variables in the formula,

$$\{x,y\} = \int \rho_{XY}(x,y) \log \left[\frac{\rho_{XY}(x,y)}{\rho_X(x)\rho_Y(y)} \right] dx \, dy \qquad (3.20)$$

gives the general idea.

Explicit use of the invariance property will not be made, but it is important to understand that this plus the uniqueness of the Information Entropy, as to its functional form, ensure that the quantity dealt with is actually a well defined physical one. Its units depend only upon the base chosen for the logarithm (natural logs here so *nats*). The limits to quantum knowledge of the state that shall be developed are therefore inherent.[4]

Having developed relevant ideas from quantum measurement theory, statistical inference and communication theory as seperate streams they can now be combined with a concise statement of the Optimal State Determination Problem.

---

[4]Assuming standard measurement theory with all observation ruled by Process One.

Chapter 4

Formulation of the OSDP

4.1 The Quantum Die

In order to emphasise that the measurement theory being discussed is influenced by the dimensionality of the state space alone let us introduce a system concept that is divorced from any particular quantum realisation. This is the Quantum Die. It is a d–faced object to which is associated a d–dimensional quantum state space ($d \times d$ density matrix: pure for a hard die, impure for a soft die). The concept is imperfect because such a die must be cast in different ways in order to express the different face probabilities associated with alternative choices of measurement basis. However, a little Demon could be called up to perform this function. If he tells us which measurement basis he chose each time then the situation described is equivalent to that of the analyser-preparator model. Such knowledge of which demonic "wrist-action" was applied in each trial is assumed.[1]

If the same wrist-action, same measurement basis, is used every time by the Demon then the Quantum Die appears as a Classical Die, albeit loaded. Selecting different wrist-actions changes the die loading. If such were not revealed to us then Demons with Quantum Dice would be hard gamblers to beat! Indeed, if the Demon's die is pure then he could always select a wrist-action to guarantee any face as up.

To each die there is a state. So Quantum Dice should be distinguishable.

---

[1]This is intended purely as a metaphor for the possibility of choosing a different basis with which to measure the die.

Two Demons could tell Quantum Dice with different states apart by throwing them repeatedly and recording the face frequencies in the same way as Mortals do with Classical Dice. The only difference is that to fully pin down the quantum state a selection of frequency data for different wrist-actions is required. For the Quantum Die realised as an analyser-preparator couple, the different wrist-actions correspond to alternative settings of the analyser that yield skewed measurement bases, they are the eigenvectors of non-commuting Hermitian operators. It matters not whether one thinks of such an abstract die or the analyser-preparator system. The essence of state determination is to evaluate a state specified by the complex components on a single basis in terms of the moduli of such components (which give probabilities) on several skewed bases.

The abstract die model is included both as an alternative way to think about finite state quantum systems and to make a connection with Bayes' original problem concerning the measurement of coin probabilities. Bayes' Rule of Inference provides a compact solution to this problem. The corresponding Quantum State Determination Problem is simply a natural generalisation of this. Describing Classical Coins and Classical Dice by states on a real Hilbert space with the natural but trivial measurement theory allowing only one measurement basis establishes the route for this generalisation.

4.2 Optimal State Determination

Combining the results of the last three chapters the Optimal State Determination Problem (hereafter $OSDP$) can now be formulated.

There are N subsystems to be analysed (N casts of a given Quantum Die). Pure preparation is assumed. So the preparator is described by some unknown state $|\psi><\psi|$. Individual subsystems can be analysed with a different measurement basis. There are to be N such orthonormal bases specified as $d \times d$ unitary matrices $\{U_k\}_{k=1}^N$ where d is the dimensionality of the state space. With respect to an arbitrary reference basis $\{|l>\}_{l=1}^d$ their matrix elements are $U_k(j,l) = <l|\phi_j^k>$. So $|\phi_j^k> = U_k|l>$ and one can identify $U_k \sim \{\phi_j^k\}_{j=1}^d$. Because overall phase is not important to calculation of outcome probabilities these unitary operators reside in equivalence classes associated with vectors in the real Euclidean space picture of density matrices

$$U_k \sim \{\phi_j^k\}_{j=1}^d \equiv \{|\phi_j^k><\phi_j^k|\}_{j=1}^d \sim V_k, \qquad (4.1)$$

42

$$V_k = \{\mathbf{v}_j^k \in \mathbf{R}^{d^2-1}\}_{j=1}^d.$$

In each trial one of d possible outcomes result. The probabilities, conditioned by the true preparator state, are

$$p(\phi_j^k|\psi) = <\psi|\phi_j^k><\phi_j^k|\psi>, \tag{4.2}$$

where

$$\sum_{j=1}^d p(\phi_j^k|\psi) = 1. \tag{4.3}$$

The kth outcome is recorded as some value for j, that is, the jth element of the kth basis was seen. This is denoted ϕ_j^k although $\phi_{j_k}^k$ is more descriptive, $j_k \in [1,d]$ is then the outcome label with $k \in [1,N]$ indexing the N trials. This expanded notation is understood in what follows. A given memory sequence of N outcomes $\{\phi_j^k\}_{k=1}^N$ is compactly denoted $\Phi_N \equiv \{\phi_j^k\}_{k=1}^N$ and has conditional probability

$$p(\Phi_N|\psi) = \prod_{k=1}^N <\psi|\phi_j^k><\phi_j^k|\psi> . \tag{4.4}$$

Having chosen an *a priori* distribution $p_0(\psi)$ over objects $|\psi><\psi|$ (matrices but also vectors), Bayesian inference yields

$$p(\psi|\Phi_N) = \frac{p(\Phi_N|\psi)p_0(\psi)}{\int p(\Phi_N|\psi)p_0(\psi)d\psi}. \tag{4.5}$$

This is the formal solution to the inversion problem. Some kind of integration over $|\psi><\psi|$ considered as continuously varying possible preparator states has to be decided upon. This is particularly simple for two state systems. Details will be explained shortly and expanded upon in the relevant sections.

Output of the inversion procedure is a probability distribution. Notice that there are a finite number of outcomes so only a finite number of outputs are possible from an N trial experiment. The outputs express the prejudice of the choice of analyser settings, $\{U_k\}_{k=1}^N$. Retention of this probability distribution is only useful if further analyser trials are to be done. Otherwise it can be reduced to the density matrix

$$\rho(\Phi_N) = \int |\psi><\psi|p(\psi|\Phi_N)d\psi. \tag{4.6}$$

43

This should be considered the final output of the state determination. It yields correct results for any expectation value calculated from the inferred distribution eq.(4.5). For any particular choice of N analyser settings there are d^N such possible density matrices as output. These are preordained by the inversion procedure eq.(4.5). They could be worked out before hand and the correct density matrix then looked up from the outcome Φ_N.

Some outputs turn out to be more *a priori* probable than others. These are good outputs corresponding to near-pure inferred density matrices for the assumed pure preparation. A non-pure inference must always result because eq.(4.6) is a convex combination of pure states. The degree of non-pureness (polarisation) reflects the quality of inference. Obviously confidence increases with N. The Bayesian formulation of state inference throws this intuitively obvious fact into sharp relief.

The *a priori* probabilities of given outcomes can be calculated from,

$$p(\Phi_N) = \int p(\Phi_N|\psi)p_0(\psi)d\psi. \tag{4.7}$$

Good outputs tend to be more *a priori* probable. However, many outputs tend to produce neighbouring density matrices. To get the total measure of a given output the prior probabilities eq.(4.7) of neighbouring density matrices eq.(4.6) have to be summed. If this is done in the appropriate manner then for large N it can be shown that the near-pure inferred outputs dominate. A class of nonsense outputs for assumed pure preparation have vanishingly small measure for large N. These correspond to permitted but extremely unlikely quantum outcomes.

The inversion problem is solved by eq.(4.5). The optimality problem $OSDP$ is underpinned by this. Identify the numerator of eq.(4.5) as a joint distribution

$$p(\psi, \Phi_N) = p(\Phi_N|\psi)p_0(\psi). \tag{4.8}$$

This determines the Correlation Information,

$$\{\psi, \Phi_N\}. \tag{4.9}$$

Recall that this measures the information gained about ψ from observation of Φ_N, averaged over all possible outcomes.

A good measure of optimality on apparatus should not depend on how the ϕ_j^k are represented. Recall that the Correlation Information is invariant

44

under one-one maps of the variables. The maps which change representation are the unitary transformations. This guides us to the correct choice of prior distribution. It should be invariant under all possible representations of ψ. That is, it should be unitary invariant. The correct choice is:

$$p_0(\psi) = \delta(1 - <\psi|\psi>), \tag{4.10}$$

$$d\psi = \prod_{l=1}^{d} dx_l dy_l, \tag{4.11}$$

$$|\psi> = \sum_{l=1}^{d} (x_l + iy_l)|l>. \tag{4.12}$$

Here $\{|l>\}_{l=1}^{d}$ is an arbitrary reference basis. A simple realisation of this occurs for two state systems as the uniform measure on the Poincaré sphere. So *for the special case $d = 2$ the above relations can be replaced by:*

$$p_0(\psi) \leftrightarrow \frac{1}{4\pi}, \tag{4.13}$$

$$d\psi \leftrightarrow d\Omega_{\hat{r}}, \tag{4.14}$$

$$|\psi><\psi| \leftrightarrow \frac{1}{2}(1 + \hat{r} \cdot \vec{\sigma}). \tag{4.15}$$

Here unit vector $\hat{r}$ is the vector representation of the state $|\psi><\psi|$, thought of as living in the real vector space of *2 × 2 Trace 1* Hermitian matrices.[2] Now $j \in [1,2]$ and, for fixed k, the basis elements can be similarly represented by two antiparallel vectors $\pm\hat{a}_k$:

$$|\phi_1^k><\phi_1^k| \leftrightarrow \frac{1}{2}(1 + \hat{a}_k \cdot \vec{\sigma}), \tag{4.16}$$

$$|\phi_2^k><\phi_2^k| \leftrightarrow \frac{1}{2}(1 - \hat{a}_k \cdot \vec{\sigma}). \tag{4.17}$$

which are antipodal points on the Poincaré sphere. A well known result enables eq.(4.4) to be rewritten in terms of the vector inner product, for example:

$$<\psi|\phi_1^k><\phi_1^k|\psi> = \frac{1}{2}(1 + \hat{a}_k \cdot \hat{r}), \tag{4.18}$$

$$<\psi|\phi_2^k><\phi_2^k|\psi> = \frac{1}{2}(1 - \hat{a}_k \cdot \hat{r}). \tag{4.19}$$

---

[2] $|\psi><\psi| \sim \hat{r} \in \mathbf{R}^3$, so we could write $|\hat{r}><\hat{r}|$.

Familiarity will be gained in the following chapter by working with eq.(4.13), where calculation is much simplified. The form eq.(4.10) is required when passing to systems of higher dimensionality. Therefore, for the present, development is continued in the most general context. All equations can be made specific in the Poincaré sphere picture by simple symbolic substitution as indicated.

Recalling the definition of Correlation Information, eq.(4.9) can be made explicit as

$$\{\psi, \Phi_N\} = \sum_{\forall \Phi_N} \int p(\psi, \Phi_N) \log \left[\frac{p(\psi, \Phi_N)}{p(\Phi_N) p_0(\psi)} \right] d\psi \qquad (4.20)$$

$$= \sum_{\forall \Phi_N} \int p(\Phi_N|\psi) p_0(\psi) \log[p(\Phi_N|\psi)] d\psi$$

$$- \sum_{\forall \Phi_N} p(\Phi_N) \log p(\Phi_N). \qquad (4.21)$$

It will be shown that the first term is constant, determined by the dimensionality d. The second term involves the prior probabilities of outcomes only. These are given by the formula eq.(4.7) and are determined by the choice of analyser setting. The optimality problem therefore involves varying the choice of the set of $U_k \sim \{\phi_j^k\}_{j=1}^d$, for fixed N, so as to maximise the second term (it is always positive) and so maximise the Correlation Information. In this way the optimal choice of analyser settings for N trials can be found. Alternatively, this is the optimal choice of N Demon wrist-actions to find the state of a Quantum Die.

4.3 Summary

Having presented the $OSDP$ in full generality, eqs.(4.4,4.5,4.6,4.7 and 4.21) can be picked out as the key ingredients. The inversion problem is solved by eq.(4.5) and eq.(4.6). The optimality problem is defined by eq.(4.21). To effect the solution, it is necessary to calculate prior probabilities from eq.(4.7), which are in turn defined through the fundamental quantum measurement rule expressed by eq.(4.4).

Attention now turns to doing this, first for two state systems where integration is simple, eq.(4.13), later for higher dimensional spaces where integration becomes more difficult, eq.(4.10), and progress limited.

Chapter 5

Two state OSDP

5.1 Sphere formulation for two states

To begin with, the formulæ of the last chapter shall be made explicit. Recall that any pure state $|\psi><\psi|$ can be represented faithfully by a unit vector $\hat{\mathbf{r}}_\psi$. Antipodal points are orthogonal pure states. So any measurement basis can also be represented by a unit vector $\hat{\mathbf{a}}$, where it is recognised that $-\hat{\mathbf{a}}$ is an equivalent representation. The two outcomes of one trial are indicated by recording a sign ± 1. Therefore the system state space is a unit sphere whilst the apparatus state space for a single trial is the set $\{-1, +1\}$. The single trial correlation is then

$$p(+1|\hat{\mathbf{r}}) = \frac{1}{2}(1 + \hat{\mathbf{a}} \cdot \hat{\mathbf{r}}), \qquad (5.1)$$

$$p(-1|\hat{\mathbf{r}}) = \frac{1}{2}(1 - \hat{\mathbf{a}} \cdot \hat{\mathbf{r}}). \qquad (5.2)$$

Since each trial may involve a different choice of analyser setting, this shall be thought of as a function parametrised by $\hat{\mathbf{a}}$. An N–trial correlation can now be formed. This is parametrised by a choice of N vectors $A_N = \{\hat{\mathbf{a}}_i\}_{i=1}^{N}$.[1] To keep track of the 2^N outcomes, introduce the set $\Phi_N = \{\sigma_i\}_{i=1}^{N}$ where $\sigma_i = \pm 1$. Then the N–trial correlation parametrised by A_N is defined by

$$p(\Phi_N|\hat{\mathbf{r}}) = \prod_{i=1}^{N} \frac{1}{2}(1 + \sigma_i \hat{\mathbf{a}}_i \cdot \hat{\mathbf{r}}). \qquad (5.3)$$

---

[1] Note that equivalent A_N are generated by inversions and permutations of the $\hat{\mathbf{a}}_i$.

47

The goal of the *OSDP* is to find a set of N unit vectors A_N which maximise the Correlation Information for this N–trial correlation.

Introduce the uniform prior $p_0(\hat{\mathbf{r}}) = 1/4\pi$. Then $p_0(\hat{\mathbf{r}})d\Omega_{\hat{\mathbf{r}}} \equiv d\hat{\Omega}_{\hat{\mathbf{r}}}$, so that the uniform prior on pure states is equivalent to a normalised surface integral on the Poincaré sphere.

Thus the equations of the last chapter become:

$$p(\Phi_N) = \int p(\Phi_N|\hat{\mathbf{r}})d\hat{\Omega}_{\hat{\mathbf{r}}} \qquad (5.4)$$

$$p(\hat{\mathbf{r}}|\Phi_N) = \frac{p(\Phi_N|\hat{\mathbf{r}})/4\pi}{\int p(\Phi_N|\hat{\mathbf{r}})d\hat{\Omega}_{\hat{\mathbf{r}}}} \qquad (5.5)$$

$$\{\hat{\mathbf{r}}, \Phi_N\} = \sum_{\forall \Phi_N = \{\sigma_i\}_{i=1}^N} \int p(\Phi_N|\hat{\mathbf{r}}) \log p(\Phi_N|\hat{\mathbf{r}})d\hat{\Omega}_{\hat{\mathbf{r}}}$$

$$- \sum_{\forall \Phi_N = \{\sigma_i\}_{i=1}^N} p(\Phi_N) \log p(\Phi_N). \qquad (5.6)$$

The inferred density matrix after N trials can be calculated from

$$\langle \hat{\mathbf{r}} \rangle = \int \hat{\mathbf{r}}\, p(\hat{\mathbf{r}}|\Phi_N)\, d\hat{\Omega}_{\hat{\mathbf{r}}}$$

as,

$$\rho(\Phi_N) = \int \frac{1}{2}(1+\hat{\mathbf{r}} \cdot \vec{\sigma})p(\hat{\mathbf{r}}|\Phi_N)\, d\hat{\Omega}_{\hat{\mathbf{r}}}$$

$$= \frac{1}{2}(1+\langle \hat{\mathbf{r}} \rangle \cdot \vec{\sigma}). \qquad (5.7)$$

This shows that $\langle \hat{\mathbf{r}} \rangle$ is a sufficient estimator for the preparator state. The probability distribution contains information for further inference, but only its mean is significant for obtaining the state after a fixed number of observations. Furthermore, the density matrix is easily seen to be necessarily mixed as the probability distribution is over a compact surface. An interior point of the Poincaré sphere must obtain and this will approach the pure surface as $p(\hat{\mathbf{r}}|\Phi_N)$ concentrates about some direction $\hat{\mathbf{r}}_0$ with increasing N.

Having fixed notation the intended sequence of calculation can now be set out. First eq.(5.6) is shown to reduce to a constant plus the Information Entropy of the prior probabilities. Knowing this it is clear that the major work consists of doing the integral in eq.(5.4). Once this is known both the Correlation Information and inferred distribution eq.(5.5) are easily calculated. Two regimes of behaviour are to be considered.

For small N eq.(5.4) yields a manageable series expansion in terms of pair inner-products of the $\hat{\mathbf{a}}_i$. Then $\{\hat{\mathbf{r}}, \Phi_N\}$ may be calculated exactly for any geometry of A_N. Optimal geometries are found for $N \leq 12$ by a computer search using the method of simulated annealing. This same series expansion yields a closed form for the output density matrices in eq.(5.7). Thus a complete solution is possible.

For large N the above method of approach still works in principle but its implementation is defeated by rapid growth in the number of terms present. However, such rapid growth proves advantageous as it enforces speedy development to an asymptotic trend. Therefore attention turns to deriving an asymptotic form for eq.(5.4). This analysis yields as an estimator of the geometry optimality a derived anisotropy matrix defined on A_N. This can be carried through in special form to eq.(5.6) and it is possible to obtain a condition for large N optimality. One special geometry is picked out and it is compared with a convenient set generated by the Platonic solids. The asymptotic form of eq.(5.7) is also obtained.

5.2 Simplification of the Correlation Information formula

Here it is shown that the first term of eq.(5.6) is a constant. Substituting for the N–trial correlation from eq.(5.3), this term becomes:

$$C_N = \sum_{\forall\{\sigma_i = \pm 1\}_{i=1}^N} \int \prod_{j=1}^N \frac{1}{2}(1+\sigma_j \hat{\mathbf{a}}_j \cdot \hat{\mathbf{r}}) \log\left[\prod_{k=1}^N \frac{1}{2}(1+\sigma_k \hat{\mathbf{a}}_k \cdot \hat{\mathbf{r}})\right] d\hat{\Omega}_{\hat{\mathbf{r}}}. \quad (5.8)$$

Expanding the log of the product as a sum yields,

$$C_N = \sum_{\forall\{\sigma_i = \pm 1\}_{i=1}^N} \int \prod_{j=1}^N \frac{1}{2}(1+\sigma_j \hat{\mathbf{a}}_j \cdot \hat{\mathbf{r}}) \left[\sum_{k=1}^N \log \frac{1}{2}(1+\sigma_k \hat{\mathbf{a}}_k \cdot \hat{\mathbf{r}})\right] d\hat{\Omega}_{\hat{\mathbf{r}}}. \quad (5.9)$$

Now think of carrying out the sum over signs outside the integral. The integral is a sum of N integrals for different fixed values of index k. As the sum over signs is carried out notice that the log argument is only changed when the sign σ_k is changed. Therefore keep this fixed and sum over all signs σ_j for $j \neq k$. Cancellation occurs for all terms in the leading j–indexed product except for that term indexed by $j = k$. It follows that the above equation reduces to

$$C_N = \sum_{k=1}^{N} \sum_{\sigma_k = \pm 1} \int \frac{1}{2}(1 + \sigma_k \hat{\mathbf{a}}_k \cdot \hat{\mathbf{r}}) \log \frac{1}{2}(1 + \sigma_k \hat{\mathbf{a}}_k \cdot \hat{\mathbf{r}}) \, d\hat{\Omega}_{\hat{\mathbf{r}}}. \tag{5.10}$$

By symmetry of the uniform sphere integration each of these integrals is the same for different $\hat{\mathbf{a}}_k$ and for both choices of sign. Thus it is only necessary to calculate the single integral

$$C_0 = \int \frac{1}{2}(1 + \hat{\mathbf{a}} \cdot \hat{\mathbf{r}}) \log \frac{1}{2}(1 + \hat{\mathbf{a}} \cdot \hat{\mathbf{r}}) \, d\hat{\Omega}_{\hat{\mathbf{r}}},$$

where

$$C_N = 2N \times C_0.$$

This is easily done if one identifies $\hat{\mathbf{a}}$ as the z axis, whence the rotational symmetry of $\hat{\mathbf{a}} \cdot \hat{\mathbf{r}}$ about this axis allows us to write

$$
\begin{aligned}
C_0 &= \frac{1}{2} \int_{-1}^{+1} \left(\frac{1+z}{2}\right) \log \left(\frac{1+z}{2}\right) dz \\
&= \int_0^1 u \log u \, du \\
&= \left[\frac{u^2}{2} \log u - \frac{u^2}{4} \right]_0^1 = -1/4.
\end{aligned} \tag{5.11}
$$

Recall that the Information Entropy of the prior probabilities is given by:

$$H(\Phi_N) = - \sum_{\forall \Phi_N} p(\Phi_N) \log p(\Phi_N).$$

Then combining the above calculation with the original form of eq.(5.6) gives the following simple expression for the Correlation Information:

$$\{\hat{\mathbf{r}}, \Phi_N\} = H(\Phi_N) - N/2, \tag{5.12}$$

showing that knowledge of the prior probabilities alone is sufficient to calculate $\{\hat{\mathbf{r}}, \Phi_N\}$. This is a considerable simplification.

Notice that $H(\Phi_N)$ is an Information Entropy on a discrete set and is always positive. Recall the Correlation Information is always non-negative.

The presence of the negative term shows that $H(\Phi_N)$ has a non-zero lower bound, which must be bigger than $N/2$. The maximum value possible for $H(\Phi_N)$ is when all $p(\Phi_N) = 2^{-N}$, the uniform distribution. So it is true that

$$0 \le \{\hat{\mathbf{r}}, \Phi_N\} \le N \log 2 - N/2.$$

Thus quantum two state measurement certainly involves constraints upon the information available about the state in N trials. These constraints shall be refined as part of the exploration of different possible geometries for A_N.

It should also be clear from the presence of $H(\Phi_N)$ that optimal choices of A_N are going to necessarily involve an approach to uniform prior probability over all possible outputs. This illustrates how an optimal apparatus is in some sense that which exhibits least prejudice. Its outputs, as indicators of the preparator state, will occur with the best available approximation to equal frequency when the average over all possible inputs is taken.[2]

Such qualitative observations form a useful backdrop for what follows. In order to make comparisons of different A_N we now develop some formulæ for $p(\Phi_N)$ and hence $H(\Phi_N)$ and $\{\hat{\mathbf{r}}, \Phi_N\}$. Two special cases are considered first. The *singlet* A_N, involving repeated trials with a single basis and the *triplet* A_N involving repeated, and equal, use of the bases $\hat{\mathbf{x}}, \hat{\mathbf{y}}$ and $\hat{\mathbf{z}}$. Then a general formula for any A_N is obtained.

5.3 Singlet probability formula

Singlet measurement involves repeated use of a single basis. Therefore the subscript on $\hat{\mathbf{a}}_i$ may be dropped and the formula for the prior outcome probabilities becomes

$$p(\Phi_N) = \int \prod_{i=1}^{N} \frac{1}{2}(1 + \sigma_i \hat{\mathbf{a}} \cdot \hat{\mathbf{r}}) \, d\hat{\Omega}_{\hat{\mathbf{r}}}. \tag{5.13}$$

For each outcome count the number of $\sigma_i = +1$. Let this be k; then the number of $\sigma_i = -1$ is $N - k$. Clearly the above equation reduces to

$$p(k) = \int \left\{ \frac{1}{2}(1 + \hat{\mathbf{a}} \cdot \hat{\mathbf{r}}) \right\}^k \left\{ \frac{1}{2}(1 - \hat{\mathbf{a}} \cdot \hat{\mathbf{r}}) \right\}^{N-k} d\hat{\Omega}_{\hat{\mathbf{r}}}.$$

---

[2]Already it should be clear that it is the geometry of state space that imposes the bounds upon $H(\Phi_N)$ and hence $\{\psi, \Phi_N\}$.

The same symmetry considerations as before apply so this can be rewritten as

$$p(k) = \frac{1}{2} \int_{-1}^{1} \left(\frac{1+z}{2}\right)^k \left(\frac{1-z}{2}\right)^{N-k} dz. \qquad (5.14)$$

The simple affine change $u = (1+z)/2$ yields

$$p(k) = \int_0^1 u^k (1-u)^{N-k} du, \qquad (5.15)$$

which can be identified as the same integral as that appearing in §2.4.2 in relation to the coin inference problem. Transcribing this result we find

$$p(k) = \left[(N+1) \binom{N}{k} \right]^{-1}. \qquad (5.16)$$

Now among the 2^N outcomes $\{\sigma_i\}_{i=1}^N$ each value k has weight $\binom{N}{k}$. Therefore $H(\Phi_N)$ can be calculated as

$$
\begin{aligned}
H(\Phi_N) &= -\sum_{k=0}^{N} \binom{N}{k} \times \left[(N+1)\binom{N}{k} \right]^{-1} \log \left[(N+1)\binom{N}{k} \right]^{-1} \\
&= +\frac{1}{N+1} \times \sum_{k=0}^{N} \left[\log(N+1) + \log \binom{N}{k} \right] \\
&= \log(N+1) + \frac{1}{N+1} \times \sum_{k=0}^{N} \log \binom{N}{k}.
\end{aligned}
\qquad (5.17)
$$

The asymptotics of this expression are worked out in Appendix A. The large N Correlation Information is found to be:

$$\{\hat{\mathbf{r}}, \Phi_N\} \sim \frac{1}{2} \log N - \frac{1}{2}(\log 2\pi - 1). \qquad (5.18)$$

5.4 Triplet probability formula

Triplet measurement involves repeated and equal use of an orthogonal triad of basis vectors, $\hat{\mathbf{x}}, \hat{\mathbf{y}}$ and $\hat{\mathbf{z}}$. These are the mutually unbiased bases of §1.3

for the case of two states. In order to make direct comparison with singlet measurement we demand that each basis is used n times, where $N = 3n$. Each outcome Φ_N is now fully parametrised by three k–values. These are k_x, k_y and k_z, the plus–counts on each of the three bases. Thus $p(\Phi_N)$ is written

$$
p(k_x, k_y, k_z) = \int \left(\frac{1+x}{2}\right)^{k_x} \left(\frac{1-x}{2}\right)^{n-k_x} \times
$$
$$
\left(\frac{1+y}{2}\right)^{k_y} \left(\frac{1-y}{2}\right)^{n-k_y} \left(\frac{1+z}{2}\right)^{k_z} \left(\frac{1-z}{2}\right)^{n-k_z} d\hat{\Omega}_{\hat{r}}, \quad (5.19)
$$

with $\hat{r} = (x, y, z)$. A series expansion is now sought in terms of the simpler integrals[3]

$$
I_{(a,b,c)} \equiv \int x^{2a} y^{2b} z^{2c} \, d\hat{\Omega}_{\hat{r}} \quad \text{where } a, b, c \in \mathbf{Z}.
$$

Take the factor in x and expand using the binomial theorem to obtain

$$
\left(\frac{1+x}{2}\right)^{k_x} \left(\frac{1-x}{2}\right)^{n-k_x} = 2^{-n} \sum_{r=0}^{k_x} \sum_{s=0}^{n-k_x} (-1)^s \binom{k_x}{r} \binom{n-k_x}{s} x^{r+s}. \quad (5.20)
$$

Now pick out the coefficient of x^{2m} for some $m \in [0, [n/2]]$. This is given by

$$
C_{2m}(k_x) = 2^{-n} \sum_{r=0}^{\min(2m, k_x)} (-1)^r \binom{k_x}{r} \binom{n-k_x}{2m-r}. \quad (5.21)
$$

So the non-zero contribution from the factor in x can be included as the power series

$$
2^{-n} \sum_{m=0}^{[n/2]} C_{2m}(k_x) x^{2m},
$$

similarly for factors in y and z. Therefore eq.(5.19) is equal to the integral of the product of three such power series. In particular it is given by

$$
p(k_x, k_y, k_z) = 2^{-N} \sum_{a=0}^{[n/2]} \sum_{b=0}^{[n/2]} \sum_{c=0}^{[n/2]} C_{2a}(k_x) C_{2b}(k_y) C_{2c}(k_z) \times I_{(a,b,c)}, \quad (5.22)
$$

---

[3]Moments with odd powers vanish through parity considerations.

with the integrals $I_{(a,b,c)}$ defined as above. These are constants and can be calculated in several ways.

Suppose a function of $\mathbf{r}$, $f(\mathbf{r})$, satisfies the scaling relation

$$f(\mathbf{r}) = r^{2k} f(\hat{\mathbf{r}}).$$

Then consider the integral

$$\int f(\mathbf{r}) \exp\{-1/2r^2\}\, d^3\mathbf{r},$$

and observe that it factorises into two parts:

$$\int f(\mathbf{r}) \exp\{-1/2r^2\}\, d^3\mathbf{r} = 4\pi \int f(\hat{\mathbf{r}})\, d\hat{\Omega}_{\hat{\mathbf{r}}} \times \int_0^\infty r^{2k+2} \exp\{-1/2r^2\}\, dr, \quad (5.23)$$

where volume elements are related by $d^3\mathbf{r} \equiv 4\pi r^2 dr\, d\hat{\Omega}_{\hat{\mathbf{r}}}$. Rearranging the above expression to solve for the surface integral yields

$$\int f(\hat{\mathbf{r}})\, d\hat{\Omega}_{\hat{\mathbf{r}}} = \frac{\int f(\mathbf{r}) \exp\{-1/2r^2\}\, d^3\mathbf{r}}{4\pi \int_0^\infty r^{2k+2} \exp\{-1/2r^2\}\, dr}. \quad (5.24)$$

Now, the integrand of $I_{(a,b,c)}$ satisfies the given scaling relation with $d = a + b + c$. Substituting $f(\mathbf{r}) = x^{2a} y^{2b} z^{2c}$ into the above shows that

$$I_{(a,b,c)} = \frac{\int_0^\infty \int_0^\infty \int_0^\infty x^{2a} y^{2b} z^{2c} \exp\{-1/2(x^2 + y^2 + z^2)\}\, dx\, dy\, dz}{4\pi \int_0^\infty r^{2k+2} \exp\{-1/2r^2\}\, dr}. \quad (5.25)$$

Noting that

$$\int_0^\infty u^{2k} \exp\{-1/2u^2\}\, du = \sqrt{\pi/2}\,(2k - 1)!!,$$

the above quotient can be calculated. After cancellation of some numerical factors it is found to be

$$I_{(a,b,c)} = \frac{(2a - 1)!!\,(2b - 1)!!\,(2c - 1)!!}{(2a + 2b + 2c + 1)!!}. \quad (5.26)$$

54

Returning to the expression eq.(5.22) we can now obtain the desired probability formula,

$$p(k_x, k_y, k_z) = 2^{-N} \sum_{a=0}^{[n/2]} \sum_{b=0}^{[n/2]} \sum_{c=0}^{[n/2]}$$
$$\left\{ C_{2a}(k_x) C_{2b}(k_y) C_{2c}(k_z) \times \frac{(2a-1)!!\,(2b-1)!!\,(2c-1)!!}{(2a+2b+2c+1)!!} \right\}. \quad (5.27)$$

This formula for triplet probabilities can be used to calculate the corresponding $H(\Phi_N)$ from

$$H(\Phi_N) = - \sum_{k_x=0}^{n} \sum_{k_y=0}^{n} \sum_{k_z=0}^{n} \binom{n}{k_x} \binom{n}{k_y} \binom{n}{k_z} \times p(k_x, k_y, k_z) \log p(k_x, k_y, k_z),$$

$$(5.28)$$

where $N = 3n$. The formula is of course exact. However, we shall only use eq.(5.28) as a convenient form for numerical calculations.

5.5 General probability formula

In this case A_N has no special symmetry and we must work with the full expression for the general correlation, eq.(5.3). This is

$$p(\{\sigma_i\}_{i=1}^N) = \int \prod_{i=1}^{N} \frac{1}{2}(1 + \sigma_i \hat{\mathbf{a}}_i \cdot \hat{\mathbf{r}})\, d\hat{\Omega}_{\hat{\mathbf{r}}}. \quad (5.29)$$

An exact series expression will be found by constructing a suitable generating function. Before attempting this eq.(5.3) needs to be broken down. This is done with the aid of some special notation.

Define an order $2K$ selection on N, as a $2K$–tuple of indices

$$\mathcal{S}_{2K} = (\tau_1, \ldots, \tau_{2K})$$

where $\tau_i \in [1, N]$ and satisifies $\tau_i < \tau_j$ for $i < j$. These $2K$–tuples have a natural ordering when considered as a $2K$–digit, base N, integer (take digits: $\tau_i - 1$). There are $\binom{N}{2K}$ possible $2K$ selections and the natural ordering provides a nice way to index them. So we write $\mathcal{S}_{2K}(J)$ where $J \in [1, \binom{N}{2K}]$.

55

With the above selection function eq.(5.29) can be expanded in a fashion analogous to that used to calculate eq.(5.19). First we define two quantities determined by a given selection $S_{2K}(J)$. These are a sign and an integral:

$$Sgn[S_{2K}(J)] = \sigma_{\tau_1(J)}\sigma_{\tau_2(J)}\cdots\sigma_{\tau_{2K}(J)} \qquad (5.30)$$

$$Int[S_{2K}(J)] = \int \prod_{i=1}^{2K}(\hat{\mathbf{a}}_{\tau_i(J)}\cdot\hat{\mathbf{r}})\,d\hat{\Omega}_{\hat{\mathbf{r}}}. \qquad (5.31)$$

With these definitions eq.(5.29) is equal to

$$p(\{\sigma_i\}_{i=1}^N) = 2^{-N}\left\{1 + \sum_{K=1}^{[N/2]}\sum_{J=1}^{\binom{N}{2K}} Sgn_{2K}[J] \times Int_{2K}[J]\right\}. \qquad (5.32)$$

As before, parity considerations ensure that only selections of even order need be taken.

The integrals $Int_{2K}[J]$ contain, as a special case, the $I_{(a,b,c)}$ encountered earlier. A different approach is required here based upon the following result.

Consider the integral

$$\int e^{\mathbf{b}\cdot\hat{\mathbf{r}}}\,d\hat{\Omega}_{\hat{\mathbf{r}}}.$$

The exponent has rotational symmetry about the direction $\hat{\mathbf{b}}$. Choosing this as the z–axis and writing $\mathbf{b}\cdot\hat{\mathbf{r}} = bz$, with $b = \|\mathbf{b}\|$ gives

$$\int e^{\mathbf{b}\cdot\hat{\mathbf{r}}}\,d\hat{\Omega}_{\hat{\mathbf{r}}} = \frac{1}{2}\int_{-1}^{+1}e^{bz}\,dz = b^{-1}\sinh b. \qquad (5.33)$$

Expanding $b^{-1}\sinh b$ as a power series yields

$$b^{-1}\sinh b = \sum_{k=0}^{\infty}\frac{b^{2k}}{(2k+1)!}. \qquad (5.34)$$

This power series will be the basis for our generating function.

It is certainly true that,

$$\int \prod_{i=1}^{2K}(\hat{\mathbf{a}}_i\cdot\hat{\mathbf{r}})\,d\hat{\Omega}_{\hat{\mathbf{r}}} = \left[\frac{\partial}{\partial\lambda_1}\cdots\frac{\partial}{\partial\lambda_{2K}}\int\exp\{\sum_{i=1}^{2K}\lambda_i\hat{\mathbf{a}}_i\cdot\hat{\mathbf{r}}\}\,d\hat{\Omega}_{\hat{\mathbf{r}}}\right]_{\vec{\lambda}=0}. \qquad (5.35)$$

Define a vector $\mathbf{b}$ by:

$$\mathbf{b}(\vec{\lambda}) \equiv \sum_{i=1}^{2K}\lambda_i\hat{\mathbf{a}}_i.$$

56

Making use of the earlier calculation of $\int e^{\mathbf{b} \cdot \hat{\mathbf{r}}} \, d\hat{\Omega}_{\hat{\mathbf{r}}}$, it is easy to see that

$$\int \prod_{i=1}^{2K} (\hat{\mathbf{a}}_i \cdot \hat{\mathbf{r}}) \, d\hat{\Omega}_{\hat{\mathbf{r}}} = \left[\frac{\partial}{\partial \lambda_1} \cdots \frac{\partial}{\partial \lambda_{2K}} \sum_{j=0}^{\infty} \frac{b(\vec{\lambda})^{2j}}{(2j+1)!} \right]_{\vec{\lambda}=0}. \qquad (5.36)$$

Notice that all powers are of even degree. Furthermore b^2 can be written in the simple form:

$$b^2(\vec{\lambda}) = \sum_{i=1}^{2K} \sum_{j=1}^{2K} \lambda_i \lambda_j (\hat{\mathbf{a}}_i \cdot \hat{\mathbf{a}}_j) = \vec{\lambda} M \vec{\lambda},$$

where $M_{ij} = \hat{\mathbf{a}}_i \cdot \hat{\mathbf{a}}_j$. Adopting for the moment the $\vec{\lambda} M \vec{\lambda}$ form the generating function is

$$\int \prod_{i=1}^{2K} (\hat{\mathbf{a}}_i \cdot \hat{\mathbf{r}}) \, d\hat{\Omega}_{\hat{\mathbf{r}}} = \left[\frac{\partial}{\partial \lambda_1} \cdots \frac{\partial}{\partial \lambda_{2K}} \sum_{j=0}^{\infty} \frac{\left(\vec{\lambda} M \vec{\lambda} \right)^j}{(2j+1)!} \right]_{\vec{\lambda}=0}. \qquad (5.37)$$

Observe that $2K$ differentiations with distinct λ_i are to be carried out. Following this, $\vec{\lambda}$ is set to zero. The powers of $\vec{\lambda} M \vec{\lambda}$ are easy to differentiate and it is clear that the only non–zero contribution from this infinite series comes from the term $(\vec{\lambda} M \vec{\lambda})^K$. Therefore we need to evaluate

$$\left[\frac{\partial}{\partial \lambda_1} \cdots \frac{\partial}{\partial \lambda_{2K}} \frac{\left(\vec{\lambda} M \vec{\lambda} \right)^K}{(2K+1)!} \right]_{\vec{\lambda}=0}. \qquad (5.38)$$

A further simplification is achieved by writing

$$\vec{\lambda} M \vec{\lambda} = \sum_{i=1}^{2K} \lambda_i^2 + 2 \sum_{(i,j):i<j}^{2K} \lambda_i \lambda_j \hat{\mathbf{a}}_i \cdot \hat{\mathbf{a}}_j,$$

and noticing that the first term breeds terms of zero contribution, whilst the second term contains all inner products of distinct pairs once. Dropping the first term does no harm so eq.(5.38) becomes:

$$\left[\frac{\partial}{\partial \lambda_1} \cdots \frac{\partial}{\partial \lambda_{2K}} 2^K \frac{\left(\sum_{(i,j):i<j}^{2K} \lambda_i \lambda_j (\hat{\mathbf{a}}_i \cdot \hat{\mathbf{a}}_j) \right)^K}{(2K+1)!} \right]_{\vec{\lambda}=0}. \qquad (5.39)$$

57

The sum of all distinct inner products is raised to power K and differentiated by $2K$ distinct λ_i. It follows that in the final result there must be a sum of products of K inner products, and in each, all of the $2K$ vectors appear once and only once. The terms of the sum are therefore characterised as corresponding one–one to partitions of the set of $2K$ vectors into K pairs. All $(2K-1)!!$ such partitions[4] are generated and notice that differentiation pulls down a factor of $K!$.

Let a given possible pairing be called a K–pair partition. Then with the substitution $2^K K!/(2K+1)! = (2K+1)!!$ eq.(5.39) yields the following expression:

$$\int \prod_{i=1}^{2K} (\hat{a}_i \cdot \hat{r}) \, d\hat{\Omega}_{\hat{r}} = \frac{1}{(2K+1)!!} \sum_{\substack{K-\text{pair} \\ \text{partitions}}} (\bullet, \bullet) \overset{K}{\cdots} (\bullet, \bullet) \qquad (5.40)$$

where $(\bullet, \bullet)$ denotes an inner product of a single pair of vectors from a certain K–pair partition. Running through all possible partitions there are $(2K-1)!!$ ways to fill in the template $(\bullet, \bullet) \overset{K}{\cdots} (\bullet, \bullet)$.[5]

Returning to eq.(5.32) we must substitute one of the above expressions for each integral $Int_{2K}[J]$. The selections $S_{2K}(J)$ define the appropriate subset of N vectors on which all of the possible K–pair partitions are to be formed. Thus eq.(5.32) becomes

$$p(\Phi_N) = \frac{1}{2^N} \left\{ 1 + \sum_{K=1}^{[N/2]} \sum_{J=1}^{\binom{N}{2K}} Sgn_{2K}[J](\Phi_N) \times \frac{Sum_{2K}[J]}{(2K+1)!!} \right\}, \qquad (5.41)$$

where

$$Sum[J] = \sum_{\substack{K-\text{pair} \\ \text{partitions}}} (\bullet, \bullet) \overset{K}{\cdots} (\bullet, \bullet). \qquad (5.42)$$

So, the sum over K picks all even order selections S_{2K} which are indexed by J. Each such selection defines a sign which depends upon Φ_N. So the

---

[4]There are $2K!$ permutations of vectors divided by $K!$ permutations of pairs divided by 2^K permutations within pairs.

[5]A sum over states is being carried out and it is interesting to observe that these terms are similar to the Wick expansion of quantum field theory, see[47], and to moment expansions that appear in the theory of Gaussian random fields, see [48, p.155]. The reason for this can be traced to the kind of symmetry involved.

signs are are written $Sgn_{2K}[J](\Phi_N)$. There is then a sum, $Sum_{2K}[J]$, over numbers determined by the various K–pair partitions of each such selection. This sum is outcome independent once a particular sign convention is chosen for the $\hat{a}_i$ as a definition of the basis A_N. The K–pair partition sums can therefore be treated as constants. The various different outcome probabilities then express themselves as linear combinations of these numbers with signs determined by the outcome.

The total number of terms to sum is:

$$Q(N) = \sum_{K=1}^{[N/2]} \binom{N}{2K} (2K-1)!! \qquad (5.43)$$

so the growth is very rapid. For instance $Q(10) = 9495$. This many terms appear in the sum for a single probability. Recall there are 2^N probabilities that need to be calculated in order to find $H(\Phi_N)$ exactly. Observing that the quantities $Sum_{2K}[J]$ are sign independent enables the computation to be reduced to that of $\sum_{K=1}^{[N/2]} \binom{N}{2K}$ such constants. Then the $p(\Phi_N)$ can be generated by summing these with appropriate signs for each outcome. An elementary symmetry halves the number of such probabilities needed. Even so one is left with $2^{N-1} \times \sum_{K=1}^{[N/2]} \binom{N}{2K}$ as an indication of the computational work required for a numerical calculation of $H(\Phi_N)$.

It is certainly clear that this series expression for $p(\Phi_N)$ is only practically useful as a calculational tool for small N. Indeed it can only be comfortably written down for $N \leq 4$. This is a useful exercise to fix ideas so we do that now.

Take as A_4 the set $\{\hat{a}, \hat{b}, \hat{c}, \hat{d}\}$. and let A_1, A_2 and A_3 be the subsets of the first one, two and three bases of A_4. Then in terms of the corresponding outcomes we have:

$$p_1 = \frac{1}{2} \qquad (5.44)$$

$$p_2 = \frac{1}{2^2}\left[1 + \frac{1}{3}\sigma_a\sigma_b(\hat{a},\hat{b})\right] \qquad (5.45)$$

$$p_3 = \frac{1}{2^3}\left[1 + \frac{1}{3}[\sigma_a\sigma_b(\hat{a},\hat{b}) + \sigma_a\sigma_c(\hat{a},\hat{c}) + \sigma_b\sigma_c(\hat{b},\hat{c})]\right] \qquad (5.46)$$

$$p_4 = \frac{1}{2^4}\left[1 + \frac{1}{3}[\sigma_a\sigma_b(\hat{a},\hat{b}) + \sigma_a\sigma_c(\hat{a},\hat{c}) + \sigma_a\sigma_d(\hat{a},\hat{d})\right.$$
$$\left. + \sigma_b\sigma_c(\hat{b},\hat{c}) + \sigma_b\sigma_d(\hat{b},\hat{d}) + \sigma_c\sigma_d(\hat{c},\hat{d})]\right]$$

$$+ \quad \frac{1}{15}\sigma_a\sigma_b\sigma_c\sigma_d[(\hat{\mathbf{a}},\hat{\mathbf{b}})(\hat{\mathbf{c}},\hat{\mathbf{d}}) + (\hat{\mathbf{a}},\hat{\mathbf{c}})(\hat{\mathbf{b}},\hat{\mathbf{d}}) + (\hat{\mathbf{a}},\hat{\mathbf{d}})(\hat{\mathbf{b}},\hat{\mathbf{c}})]\Big] \quad (5.47)$$

The pattern of development is evident from these first few formulæ. Also one can make an immediate rigorous conclusion about the solution of the $OSDP$ for $N \leq 3$. Recall that

$$\{\hat{\mathbf{r}}, \Phi_N\} = H(\Phi_N) - N/2. \qquad (5.48)$$

Now we know from the elementary properties of the Information Entropy that

$$0 \leq H(\Phi_N) \leq N \log 2$$

with equality at the upper bound attained only when $H(\Phi_N) = 2^{-N}$, for all Φ_N. By inspection it is clear that optimal measurement for p_2 and p_3 is achieved by using a mutually orthogonal set of $\hat{\mathbf{a}}_i$ ($N = 2$ or 3). This is not possible for $N \geq 4$ because the dimensionality of state space, $D = 2^{d^2-1} = 3$, does not allow any more such mutually orthogonal bases. As expected the choice of uniform prior, $p_0(\hat{\mathbf{r}}) = 1/4\pi$, ensures that all choices are optimal for p_1. Dependence of $p(\Phi_N)$, and therefore $H(\Phi_N)$, upon inner products alone also demonstrates that the overall orientation of a geometry A_N is unimportant.[6]

The above solution for $N \leq 3$ supports geometric intuition about how best to sample space so as to have an apparatus that is best in some average sense. Beyond $N = 3$ it is impossible to attain the desired equality of outcome probabilities. This indicates a lowering of the upper bound $N \log 2 - N/2$. Whatever this new bound is, it is best characterised asymptotically. We do this later. For the moment we have a formula that can be implemented for small N and it makes sense to engage in some numerical experiments to optimise A_N for as large an N as is practicable. The limit in terms of computational expenditure and quality of answers is about 12.

---

[6]So the choice of axes for the Poincaré sphere makes no difference. Of course this must be so to have a representation independent measure of apparatus optimality. Changing the prior to some peaked distribution implies we have initial information about some preferred direction and then the orientation of A_N with respect to this does matter.

5.6 Numerical experiments

Given the general probability formula of the last section and those for the special cases of singlet and triplet measurement, it is now possible to explore the dependence of the Correlation Information upon the chosen geometry A_N.

5.6.1 An initial guess: the Platonic solids

Outside the context of the problem as one of optimal measurement it is interesting to pursue this as a purely geometric problem. One wonders what configuration of vectors the optimisation of $H(\Phi_N)$ will choose.

The Platonic solids are suggestive as solutions to the problem in that they represent the best available isotropy for a sampling of fixed directions. There are five such regular solids: cube, tetrahedron, octahedron, dodecahedron and icosahedron. Their vertex and face directions are candidates for forming A_N. However, the possibilities are reduced from the whole set in that A_N is invariant under inversions of any component vector. Thus the cube and octahedron are equivalent, yielding the orthogonal triplet. We have already seen that this is the solution for $N = 3$. The tetrahedron vertex directions define a likely candidate for $N = 4$. The dodecahedron and icosahedron are dual figures, taking the face directions on one is equivalent to the taking the vertex directions on the other. Noting inversion symmetry, and choosing face directions, generates candidates for $N = 6$ and $N = 10$ respectively. One wonders what happens for cases such as $N = 5$ or $N = 7$ where there is no obvious choice.

5.6.2 Rationale for a computer search

It would be nice if it were possible to gain some information about the likely symmetry from the formula itself. However, this is complicated by the fact that we need to evaluate a sum like

$$- \sum_{\forall \Phi_N} p(\Phi_N) \log p(\Phi_N).$$

Without a closed form for $H(\Phi_N)$ it is difficult to make progress in this direction. There is one powerful property that the expression possesses and

61

that is convexity. This has consequences for optimisation problems but we have not found a way to exploit it.[7]

In the absence of a direct method a computer optimisation is explored. The basic idea is to pick configurations at random and retain those that represent an improvement over previous stored configurations. The only complications are coding the formula for probabilities and coping with rapidly slowing convergence as N increases. The latter problem enforces the restriction to $N \leq 12$ for a general configuration. The simpler formulæ for singlet and triplet measurement can be taken out to $N = 300$ and $N = 75$ respectively. This is done to make comparison with later asymptotic expressions.

5.6.3   Calculation algorithm

The formula for $p(\Phi_N)$ is unwieldy to code explicitly but has sufficient structure to allow us to generate it with a suitable algorithm.

The selections $\mathcal{S}_{2K}(J)$ are stored to save time. For each new A_N, we make all possible selections in sequence. From this, all possible K-pair partitions are generated from a simple recursive swapping algorithm acting on the $2K$ elements of a given selection. The numbers $Sum_{2K}[J]$ are then stored. Having accumulated this information it is a simple matter to run through all the outcomes and calculate the $p(\Phi_N)$ whilst accumulating a value for $H(\Phi_N)$ concurrently. In fact, one need only generate half of the outcomes, as changing the sign of $\hat{r}$ in the defining formula shows that a given outcome has the same probability as its inversion (reverse all signs).

There will be further such symmetries for special configurations. The triplet and singlet formulæ are examples of this. However, since such symmetries do not apply in general no attempt is made to exploit them.

---

[7]There is a nice closed form for the *variance* of the 2^N numbers $p(\Phi_N)$. The quantities $Sgn_{2K}[J](\Phi_N)$ have the orthogonality property

$$\sum_{\forall \Phi_N} Sgn_{2K}[J](\Phi_N) Sgn_{2K'}[J'](\Phi_N) = 2^{+N} \delta_{JJ'} \delta_{KK'}.$$

This ensures that

$$\langle (p - \overline{p})^2 \rangle = 2^{-N} \sum_{K=1}^{[N/2]} \sum_{J=1}^{\binom{N}{2K}} \left(\frac{Sum_{2K}[J]}{(2K+1)!!} \right)^2.$$

However, the two optimisation problems are not the same.

5.6.4 Optimisation algorithm

Simulated annealing is a very robust optimisation method that has been described in detail elsewhere[49, p.326]. Here a simple description should suffice.

To each geometry A_N, one could associate a phase space which is the Cartesian product of N spheres. To each point in this phase space there is a value for the Correlation Information. Maxima and minima are sought. These are found by taking a random walk on all spheres simultaneously. Again obvious symmetries are ignored, but we introduce the constraint that one vector be fixed and that a second be confined to a single great circle. This helps reduce the dimensionality of the random walk and so improve convergence.

At each step a random neighbouring configuration is generated and its Correlation Information is calculated and compared with the last stored value. The new value and configuration are kept with a probability related to the improvement represented by the new value. Included is a finite probability of retaining a worse configuration, so as to help get out of local minima.

The parameters that control the keep probability and possible range of exploration are made to obey an exponential type "cooling law". There is a physical analogy whereby the system undergoes diminishing thermal fluctuations whilst slowly finding its state of minimum energy. Taking this energy to be minus (or plus) the Correlation Information, defined for each A_N, enables best (or worst) geometries to be found.

5.6.5 Numerically generated optimal configurations

The above algorithm was implemented on a *VAX* system in double precision arithmetic. Random initial configurations were generated and various combinations of the simulated annealing parameters were explored.

The numerical accuracy of the routines was checked in two ways. For fixed N, there are 2^N separate probabilities that are calculated to find the Correlation Information. These must satisfy a normalisation constraint. Keeping a note of this sum showed that the probabilites were correctly normalised to twelve figures accuracy. In addition the general program, involving algorithmic generation of the probability formula, was checked against the corresponding exact formulæ available for the singlet and triplet configurations. These are much simpler to program and so provide a good check on the arithmetic.

Initially, the program was run many times through several thousand configurations from a random starting point. In this way a picture could be built up of the stability of the optimisation and of the types of symmetry that are preferred. It was possible to vary the number of configurations tested and both the initial and final values of the simulated annealing parameters. The freedom was retained to feed back the final configuration of some run as the starting configuration for subsequent runs. In this manner it was possible to monitor the progress of the optimisation and at the same time break–up the calculation into smaller parts. This is important as there is no very good stopping criterion for this problem.

Results obtained from such experimentation are given in table(5.1). There is a scale of confidence attached to these results. The optimised geometries are well located for $N \leq 6$. There is less confidence for $N = 7$ and 8, whilst those beyond $N = 8$ have not been optimised. They are simply guessed configurations included for comparison and to test ideas about likely behaviour where several Platonic solids provide alternative high symmetry configurations.

Adapting the program to search for worst configurations always yielded the repeated singlet within a short number of cycles (N–dependent but generally under 1000). This is not surprising as such a measurement scheme discards all information about the location of the state on a ring of latitude on the Poincaré sphere. Choosing a single basis only gives information about the projection of the state upon the sphere axis made by the basis direction. This results in one–dimensional localisation upon the two–dimensional sphere surface. In this simple case convergence was rapid and so there is very good confidence in this result.

Tabulated results are graphed in figure(5.1). The worst geometries provided by single basis measurement form a lower bound to what may be termed the experimental region. Running the program to search for best configurations defines an upper bound to this experimental region. It should be clear from figure(5.1) that there is an apparent smooth development in the results. Attention now turns to a case study for different N.

Reassuringly, simulated annealing reproduces the rigo rously known results for $N = 2$ and $N = 3$. This provides a check that the optimisation algorithm functions as expected.

64

N	Max. C.I. (*nats*)	A_N and repetitions		Min. C.I. (*nats*)
1	0.193 147 181	singlet	1	0.193 147 181
2	0.386 294 361	⊥ pair	1	0.329 661 349
3	0.579 441 542	⊥ triplet	1	0.435 600 505
4	0.727 479 328	reg. tet.	1	0.522 307 551
5	0.861 474 306	irreg. quintet	1	0.595 767 138
6	0.984 685 304	⊥ triplet	2	0.659 531 808
	0.982 068 934	reg. dod.	1	
7	1.094 115 112	irreg. septet.	1	0.715 886 704
8	1.195 145 276	irreg. octet	1	0.766 390 214
	1.195 090 212	reg. tet.	2	
9	1.288 891 320	⊥ triplet	3	0.812 153 537
10	1.374 724 677	reg. icos.	1	0.853 997 346
	1.374 700 181	irreg. quintet	2	
12	1.530 853 409	reg. dod.	2	0.892 545 229
	1.530 375 257	reg. tet.	3	
	1.528 748 618	⊥ triplet	4	

Table 5.1: Configurations of maximal and minimal Correlation Information. The minimal geometry is always the repeated singlet. Multiple entries in the maximal column have the best listed first with alternative high symmetry configurations included for comparison. The coordinates for these configurations are listed according to the above identifiers in Appendix B.

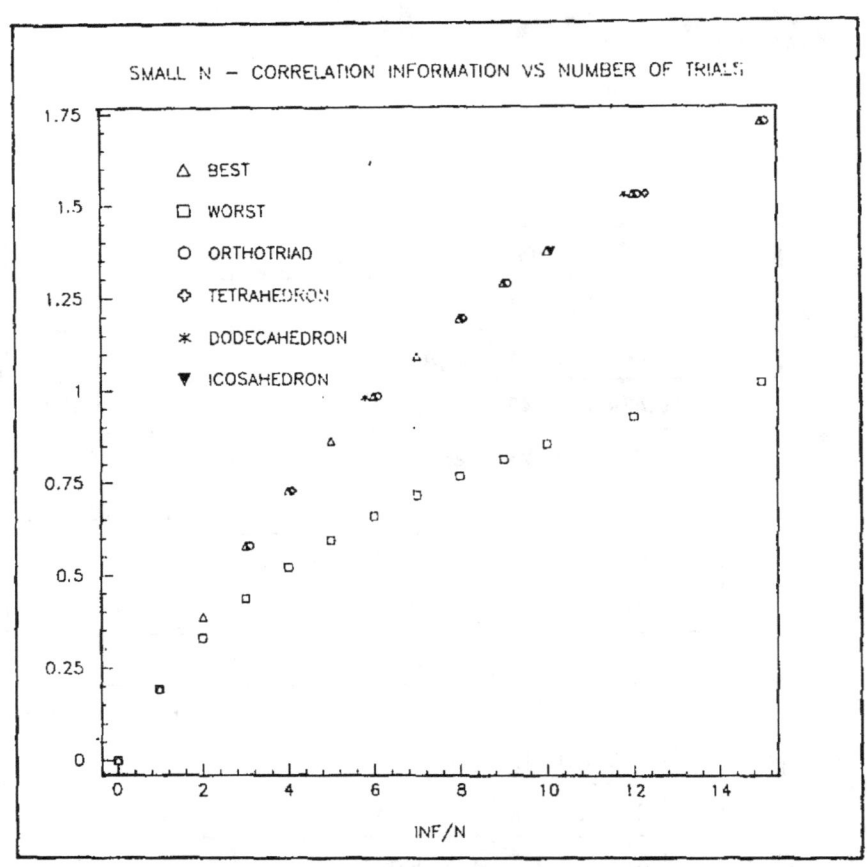

Figure 5.1: Correlation Information in nats versus N for optimal geometries as generated by a computer search. The upper curve includes both the best configuration found and other high symmetry ones as appropriate. For clarity points lying at the same N have been separated slightly. The experimental region is that area bounded by best and worst curves.

For $N = 4$ a tetrahedral configuration was found which proved to be exceptionally stable with respect to perturbation when this was used as a starting point. The initial guess concerning the Platonic solids is borne out in this case. One can understand this in terms of the fact that such a symmetry ensures that there are only two possible probability values. This represents an approach to the case of all equal probabilities, known to be the solution to the unconstrained problem of maximising the entropy.

At $N = 5$ the convergence begins to be degraded. Initial experimentation suggested a configuration where four vectors adopt a regular square pyramidal shape with the fifth pointing perpendicular to its base. Having observed this, the shape was parametrised by the angle made between the fifth vector and any of the other four. Simulated annealing was then continued with only this angle varied. The configuration found via this method is given in table(B.1). It has an optimal opening angle of $\theta \simeq 113.7^0$. Subsequently this configuration was tested for stability by relaxing the movement constraint and allowing it to try and seek a better value. No such better value was found under any starting condition.

For $N = 6$ one might have expected the dodecahedron to provide the optimal configuration. In fact this was marginally worse than that obtained by using the orthogonal triplet twice. This can be understood by recognising that the repeated triplet introduces many zero terms into the sum from which the probabilities are generated. It is interesting to note that this strategy wins over the higher symmetry provided by a dodecahedral configuration.

For $N = 7$ a particular symmetry seemed to recur in many trials. This consisted of a propensity for three groups of vectors to pair off together and so form a kind of squashed tetrahedral symmetry for an effective set of four directions. The unpaired direction approached an equiangular configuration with respect to the paired directions. This bears a strong similarity with the situation found for $N = 5$, although the doubling strategy identified for $N = 6$ appears as well. Again the suggested symmetry was enforced and the optimisation continued in terms of a changing opening angle. The search then became one dimensional with a best opening angle of $\theta \simeq 118.3^0$, see table(B.2). The dependence of Correlation Information upon opening angle is depicted in figure(5.2).

On reaching $N = 8$ a similar dimensionality-reducing procedure was attempted, however with less success. The figure represented in table(B.3) has the highest Correlation Information of any found. We note that it is

67

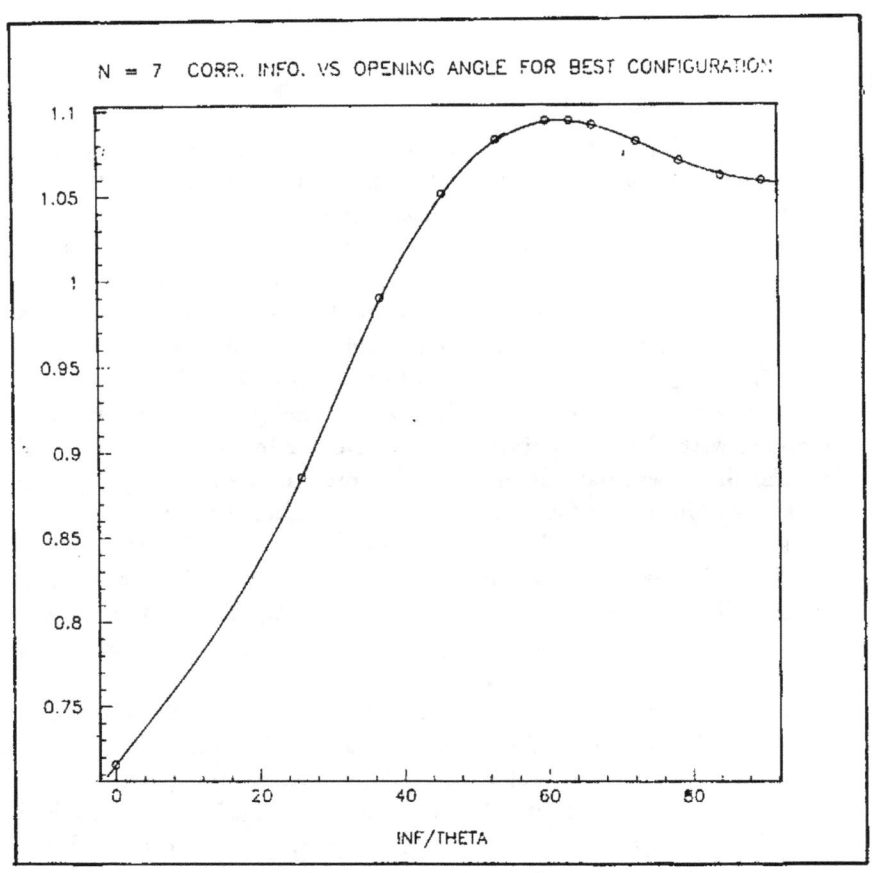

Figure 5.2: The optimal A_7 geometry parametrised by the angle, θ, between the single unrepeated direction and the other three. Hence, $\theta = 0^0$ corresponds to the singlet and $\theta = 90^0$ puts the three doubled directions in a plane with the fourth perpendicular to this. Notice that inversion symmetry ensures that θ is equivalent to $180^0 - \theta$, which produces the local minimum at $\theta = 90^0$.

certainly better than the naive guess provided by repeating the tetrahedral configuration for A_4. Further iterations of the annealing procedure for some $20,000$ configurations with a small exploration range leave the Correlation Information unchanged to nine figures.

Beyond this point the cost of the computation in relation to quality of answers led us to abandon further attempts at optimisation. For $N = 9$ and beyond, attention concentrated upon simply calculating the Correlation Information for some representative geometries. Having identified the trend for a twin strategy of repeating bases and seeking symmetric configurations, meaningful results can still be obtained by simply comparing various guessed configurations.

Calculations were carried out as far as $N = 12$ where there is the interesting possibility of comparing the repeated triplet, tetrahedron and dodecahedron. In accord with intuition these fall into the expected heirarchy, where the doubled dodecahedron is best and the four-fold repeated triplet worst. Thus the situation applying to $N = 6$ is reversed. However, note that there is little difference between these three alternatives.

Any further experimentation with this is of no practical consequence. The differences in Correlation Information are minute compared with the gains possible by making one further observation. The problem is of purely geometric interest and given the propensity for symmetry in the results it would be better attacked by analytic methods. These are available for the asymptotic case so the question is readdressed there.

The practically useful conclusion is that picking a repetition of one of the geometries provided by the Platonic solids will come close to the sought after optimality. Given that each measurement basis must be realised by a piece of physical apparatus it is worthwhile to know that choosing three such pieces results in only marginally worse performance than the optimal possible. This is especially nice given that a minimum of three non-commuting bases are required to span the space of states.

Later the estimate of the relative merits of repeated geometries, A_N, for large N will be refined through an asymptotic analysis. This enables the gap to be closed concerning the upper bound to the Correlation Information and a solution to the optimality problem as intimated above. It also provides a nice method to display the relative anisotropy of spatial sampling for various repeated geometries.

As a prelude to the asymptotics we include a graph, figure(5.3) of the

Correlation Information calculated from the exact singlet and triplet formulæ for $N < 75$. An asymptotic trend to

$$k \log N + c$$

is indicated. Earlier we calculated the asymptotic singlet constants as

$$k = 1/2 \ \text{ and } \ c = 1/2(\log 2\pi - 1).$$

From looking at the triplet curve one may conjecture that

$$k \simeq 1 \ \text{ and } \ c \simeq -1.$$

These values will be confirmed via asymptotic analysis in the next section.

5.7 Asymptotic Correlation Information for repeated measurement sets

Here we attack the problem of asymptotic Correlation Information for a special class of measurements that admit a convenient solution. These involve repeated observation with the measurement set $A_m = \{\hat{\mathbf{a}}_i\}_{i=1}^m$, where the vectors must span the space, so $m \geq 3$. Each basis is to be used n times and this parameter is taken to be large. The total number of trials is therefore $N = nm$, where m is considered fixed.

5.7.1 Outline of calculation

The analysis is based upon the following observation. If the plus count on the ith basis is denoted k_i, with $k_i \in [0, n]$, then upon forming the new outcome parameter

$$X_i = 2k_i/n - 1,$$

we can write the N trial correlation as:

$$
\begin{aligned}
p(\Phi_N | \hat{\mathbf{r}}) &\equiv p(\vec{X} | \hat{\mathbf{r}}) \\
&= \prod_{i=1}^m \left\{ \frac{1}{2}(1 + \hat{\mathbf{a}}_i \cdot \hat{\mathbf{r}}) \right\}^{\frac{n}{2}(1 + X_i)} \left\{ \frac{1}{2}(1 - \hat{\mathbf{a}}_i \cdot \hat{\mathbf{r}}) \right\}^{\frac{n}{2}(1 - X_i)}.
\end{aligned}
\tag{5.49}
$$

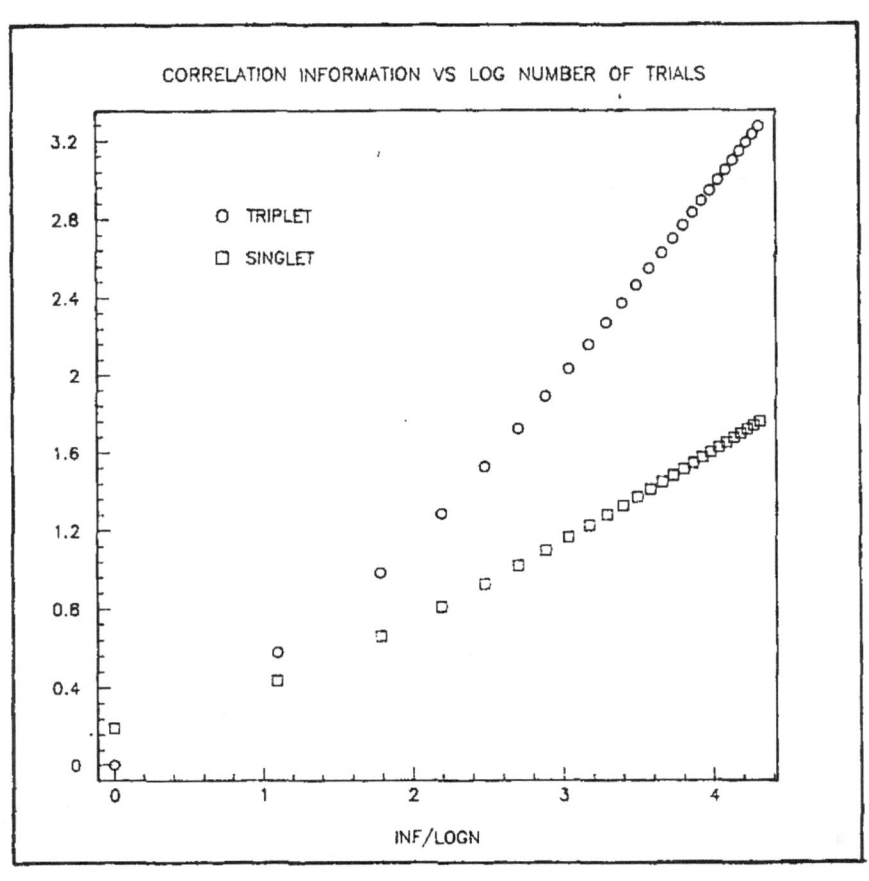

Figure 5.3: Correlation Information in natural units versus log N.

71

Here, $\vec{X} = (X_1, \ldots, X_m)$ and $X_i \in [-1, 1]$. Notice that the discrete parameters X_i take values upon a lattice of spacing $2/n$, and so simulate a continuous interval as n becomes very large. In addition each such probability has an $\vec{X}$ dependent weight given by the combinatorial factor

$$w(\vec{X}) = \prod_{i=1}^{m} \binom{n}{\frac{n}{2}(1 + X_i)}. \tag{5.50}$$

Thus a sum of the form

$$\sum_{\Phi_N} f(\Phi_N)$$

becomes

$$\sum_{\vec{X}} w(\vec{X}) f(\vec{X}).$$

This summation is identical since any $f(\Phi_N)$ can be written as $f(\vec{X})$ for some $\vec{X}$. The sum over vectors is in effect a sum over all lattice points in an m-cube formed from the Cartesian product of the m intervals $[-1, 1]$. The sum will remain a discrete sum although we shall exploit the increasing density of points in order to treat the argument of $f(\vec{X})$ as a continuous quantity.

We now calculate the inferred distribution for $\hat{\mathbf{r}}$ parametrised by observed data $\vec{X}$. In the current notation this is simply

$$p(\hat{\mathbf{r}}|\vec{X}) = \frac{p(\vec{X}|\hat{\mathbf{r}})}{\int p(\vec{X}|\hat{\mathbf{r}}) \, d\Omega_{\hat{\mathbf{r}}}}. \tag{5.51}$$

For later convenience we define

$$G(\vec{X}, \hat{\mathbf{r}}) \equiv p(\hat{\mathbf{r}}|\vec{X})$$

and observe that this function is simply proportional to $p(\vec{X}|\hat{\mathbf{r}})$ with constant chosen so that

$$\int G(\vec{X}, \hat{\mathbf{r}}) \, d\Omega_{\hat{\mathbf{r}}} = 1.$$

The normalisation constant is difficult to calculate unless $\vec{X}$ is such that $X_i = \hat{\mathbf{a}}_i \cdot \hat{\mathbf{r}}$ for some $\hat{\mathbf{r}}$. We shall defer discussion of the reason why, in order to show that this restriction is not an important difficulty.

To this end we now develop a useful form for the Correlation Information in terms of $\log G(\vec{X}, \hat{\mathbf{r}})$. Recall the symmetric expression

$$\{\hat{\mathbf{r}}, \Phi_N\} = \sum_{\Phi_N} \int p(\hat{\mathbf{r}}, \Phi_N) \log \left[\frac{p(\hat{\mathbf{r}}, \Phi_N)}{p(\Phi_N) p_0(\hat{\mathbf{r}})} \right] d\Omega_{\hat{\mathbf{r}}}. \tag{5.52}$$

The order of summation and integration is open to change. For reasons that shall become clear we elect to do the sphere integration last. Then use of Bayes' rule with uniform prior enables the Correlation Information to be recast exactly as

$$\{\hat{\mathbf{r}}, \Phi_N\} = \int \left[\sum_{\vec{X}} w(\vec{X}) p(\vec{X}|\hat{\mathbf{r}}) \log G(\vec{X}, \hat{\mathbf{r}}) \right] d\hat{\Omega}_{\hat{\mathbf{r}}}. \tag{5.53}$$

Notice that the prefactor for $\log G(\vec{X}, \hat{\mathbf{r}})$ consists of the product of m, binomial factors. A generic one is:

$$B_{\hat{\mathbf{a}}}(n, X; \hat{\mathbf{r}}) \equiv \binom{n}{\frac{n}{2}(1+X)} \left\{ \frac{1 + \hat{\mathbf{a}} \cdot \hat{\mathbf{r}}}{2} \right\}^{\frac{n}{2}(1+X)} \left\{ \frac{1 - \hat{\mathbf{a}} \cdot \hat{\mathbf{r}}}{2} \right\}^{\frac{n}{2}(1-X)}. \tag{5.54}$$

The terms $n(1 \pm X)/2$ take integer values. Under summation over $X \in [-1, 1]$, these factors have useful moment properties. In particular, define the qth central moment, ν_q by:

$$\nu_q \equiv \sum_X (X - \hat{\mathbf{a}} \cdot \hat{\mathbf{r}})^q B_{\hat{\mathbf{a}}}(n, X; \hat{\mathbf{r}}). \tag{5.55}$$

Reverting briefly to summation notation in terms of k, this becomes

$$\nu_q = \sum_{k=0}^{n} (2/n)^q \binom{n}{k} (k - nP)^q P^k (1-P)^{n-k}, \tag{5.56}$$

where $P = (1 + \hat{\mathbf{a}} \cdot \hat{\mathbf{r}})/2$. This may be identified as the usual central moment μ_q scaled by $(2/n)^q$. Only the first three shall be required; they are listed in [31, p.626]. Appropriately scaled, the μ_q become the ν_q:

$$\begin{aligned}
\nu_0 &= 1, \\
\nu_1 &= 0, \\
\nu_2 &= \left(1 - (\hat{\mathbf{a}} \cdot \hat{\mathbf{r}})^2 \right) / n.
\end{aligned} \tag{5.57}$$

73

where the first expresses normalisation.

Returning to eq.(5.53) and substituting for $w(\vec{X})p(\vec{X}|\hat{\mathbf{r}})$ gives

$$\{\hat{\mathbf{r}}, \Phi_N\} = \int \sum_{X_1} \cdots \sum_{X_m} \left[\prod_{i=1}^{m} B_{\hat{\mathbf{a}}_i}(n, X; \hat{\mathbf{r}}) \right] \log G(\vec{X}, \hat{\mathbf{r}}) \, d\hat{\Omega}_{\hat{\mathbf{r}}}. \qquad (5.58)$$

If we now observe that for large n each binomial factor is strongly peaked about $X_i = \hat{\mathbf{a}}_i \cdot \hat{\mathbf{r}}$, then it makes sense to approximate $\log G(\vec{X}, \hat{\mathbf{r}})$ by its Taylor expansion about these points. In particular, replace this by the series

$$
\begin{aligned}
\log G(\vec{X}, \hat{\mathbf{r}}) \quad \simeq \quad & \log G(\vec{X}_0, \hat{\mathbf{r}}) \\
& + \quad \sum_{i=1}^{m} \frac{\partial}{\partial X_i} \left[\log G(\vec{X}, \hat{\mathbf{r}}) \right]_{\vec{X}_0} \\
& \qquad \times (X_i - \hat{\mathbf{a}}_i \cdot \hat{\mathbf{r}}) \\
& + \frac{1}{2} \sum_{i,j=1}^{m} \frac{\partial}{\partial X_i} \frac{\partial}{\partial X_j} \left[\log G(\vec{X}, \hat{\mathbf{r}}) \right]_{\vec{X}_0} \\
& \qquad \times (X_i - \hat{\mathbf{a}}_i \cdot \hat{\mathbf{r}})(X_j - \hat{\mathbf{a}}_j \cdot \hat{\mathbf{r}}), \qquad (5.59)
\end{aligned}
$$

where

$$\vec{X}_0 = (\hat{\mathbf{a}}_1 \cdot \hat{\mathbf{r}}, \ldots, \hat{\mathbf{a}}_m \cdot \hat{\mathbf{r}}).$$

Note that such $\vec{X}_0$ certainly exist since they can be generated by taking all $\hat{\mathbf{r}}$ in the above expression.[8] In what follows $\hat{\mathbf{r}}$ is considered fixed and for each value there is a certain unique $\vec{X}$ which satisfies this condition.

In order to simplify this expression we make use of the moments listed in eq.(5.57). Notice that all terms involving first partial derivatives vanish, as do those involving second partial derivatives with $i \neq j$. Therefore, we must calculate the quantity

$$
\begin{aligned}
\{\hat{\mathbf{r}}, \Phi_N\} \quad = \quad & \int \log G(\vec{X}_0, \hat{\mathbf{r}}) \, d\hat{\Omega}_{\hat{\mathbf{r}}} \\
& + \int g(\hat{\mathbf{r}}) \, d\hat{\Omega}_{\hat{\mathbf{r}}}. \qquad (5.60)
\end{aligned}
$$

where

$$g(\hat{\mathbf{r}}) \equiv \frac{1}{2n} \sum_{i=1}^{m} \left(1 - (\hat{\mathbf{a}}_i \cdot \hat{\mathbf{r}})^2 \right) \frac{\partial^2}{\partial X_i^2} \left[\log G(\vec{X}, \hat{\mathbf{r}}) \right]_{\vec{X}_0}.$$

---

[8]This is no surprise. It expresses large n convergence of X_i to the expectation value of the corresponding operator: $\hat{M} \equiv |\hat{\mathbf{a}}_i\rangle\langle\hat{\mathbf{a}}_i| - |-\hat{\mathbf{a}}_i\rangle\langle-\hat{\mathbf{a}}_i| = \hat{\mathbf{a}}_i \cdot \vec{\sigma}$.

The first term is simple enough and can be obtained readily via other routes. The second one causes some minor trouble but can be dealt with once a proper understanding of the normalisation integration of G is gained. Then it is possible to show that it is equal to minus one and so independent of A_m.

Tedious extension of the above method shows that later terms tend to zero asymptotically. The importance of this is that the A_m dependence of the asymptotic Correlation Information is thus seen to be buried entirely in the first term. This will enable us to derive an analytic expression for the measure of A_m optimality. Later, application is made to analysis of the repeated bases derived from the Platonic solids. The special form of the quantity also suggests that the constraint of repeated measurement sets can be removed.

We stress the above in order to indicate that

$$\int \log G(\vec{X}_0, \hat{\mathbf{r}}) \, d\hat{\Omega}_{\hat{\mathbf{r}}}$$

is the major prize. The other quantity is of little interest, although it is required to obtain agreement with numerical experiments.

5.7.2 Calculation of G

First recall the expression for $p(\vec{X}|\hat{\mathbf{r}})$ given in eq.(5.49). We choose to write this in terms of the new function:

$$
F(\vec{X}, \hat{\mathbf{r}}) \equiv \sum_{i=1}^{m} \frac{1}{m} \left\{ \frac{1}{2}(1 + X_i) \log \frac{1}{2}(1 + \hat{\mathbf{a}}_i \cdot \hat{\mathbf{r}}) \right.
$$
$$
\left. + \frac{1}{2}(1 - X_i) \log \frac{1}{2}(1 - \hat{\mathbf{a}}_i \cdot \hat{\mathbf{r}}) \right\}.
$$

The factor $1/m$ is included so as to ensure that $F(\vec{X}, \hat{\mathbf{r}})$ has a uniform upper bound, irrespective of the size of m. Introducing $N = nm$ gives

$$p(\vec{X}|\hat{\mathbf{r}}) = \exp N \left\{ F(\vec{X}, \hat{\mathbf{r}}) \right\} \qquad (5.61)$$

$$G(\vec{X}, \hat{\mathbf{r}}) = \frac{1}{\mathcal{N}(\vec{X})} \exp N \left\{ F(\vec{X}, \hat{\mathbf{r}}) \right\}, \qquad (5.62)$$

with the normalisation given by,

$$\mathcal{N}(\vec{X}) = \int \exp N \left\{ F(\vec{X}, \hat{\mathbf{s}}) \right\} \, d\Omega_{\hat{\mathbf{s}}}. \qquad (5.63)$$

75

Turning to the calculation of this normalisation we shall meet the difficulty alluded to previously. This has to do with those $\vec{X}$ where it can be nicely calculated. It is at this point that the natural constraint $X_i = \hat{a}_i \cdot \hat{r}$ for some $\hat{r}$ proves useful. Such outcomes will be referred to as the $\vec{X}_0$–case.

Observe that the integrand is a function over a closed surface and so has a maximum, call this $\hat{s}_0$. The presence of the n factor in the exponent ensures that maxima become highly peaked as this parameter becomes large. The situation is similar to those encountered in standard one–dimensional asymptotic analysis[46, p.273]. However, here the integration region is curved and two–dimensional.

Nevertheless, we can still replace $F(\vec{X}, \hat{s})$ by its Taylor series around the maximum. Choosing convenient axes with $\hat{s}_0 = \hat{z}$ as the position of the maximum, we shall replace the sphere integral by that over a small paraboloidal cap with points parametrised by those in the tangent xy–plane. This takes care of the curvature and allows extension of integration limits in the usual way.

In this manner it is found that the curvature introduces a correction that disappears for the $\vec{X}_0$ case. Having discovered this it is reasonable to employ a naive tangent plane integration for what follows. At this first step it is not necessary to know what the stationary point is. It is given by an implicit function parametrised by $\vec{X}$. This is only easily solved for the constrained $\vec{X}_0$ outcomes.

Relaxing the unit vector constraint upon $\hat{s}$, we calculate a local series expansion about $\hat{s}_0$ for the function $F(\vec{X}, s)$. For this the gradient vector and Hessian matrix are required. These are easily verified to be:

$$
\begin{aligned}
\mathbf{b}(\vec{X}, s) &\equiv \nabla_s F(\vec{X}, s) \\
&= \sum_{i=1}^{m} \frac{\hat{a}_i}{m} \left\{ \frac{(1 + X_i)}{1 + \hat{a}_i \cdot s} - \frac{(1 - X_i)}{1 - \hat{a}_i \cdot s} \right\} \\
&= \sum_{i=1}^{m} \frac{\hat{a}_i}{m} \left\{ \frac{2(X_i - \hat{a}_i \cdot s)}{1 - (\hat{a}_i \cdot s)^2} \right\}
\end{aligned}
\tag{5.64}
$$

and

$$
\begin{aligned}
\mathbf{A}(\vec{X}, s) &\equiv -\nabla_s \nabla_s F(\vec{X}, s) \\
&= +\sum_{i=1}^{m} \frac{\hat{a}_i \hat{a}_i}{m} \left\{ \frac{(1 + X_i)}{(1 + \hat{a}_i \cdot s)^2} + \frac{(1 - X_i)}{(1 - \hat{a}_i \cdot s)^2} \right\}
\end{aligned}
$$

76

$$= + \sum_{i=1}^{m} \frac{\hat{\mathbf{a}}_i \hat{\mathbf{a}}_i}{m} \left\{ \frac{2(1 - 2X_i \hat{\mathbf{a}}_i \cdot \mathbf{s} + (\hat{\mathbf{a}}_i \cdot \mathbf{s})^2)}{(1 - (\hat{\mathbf{a}}_i \cdot \mathbf{s})^2)^2} \right\}. \qquad (5.65)$$

By juxtaposition of two vector quantities is meant the matrix $\mathbf{aa}^T$. Also, we have elected to define $\mathbf{A}$ as minus the Hessian for later convenience. Note that $\hat{\mathbf{s}}$ which are poles of the above do not cause trouble since it is the maxima that are of interest. Tracing back to the original expression for $p(\vec{X}|\hat{\mathbf{r}})$ such points are zeros.

The stationary points of $F(\vec{X}, \mathbf{s})$ restricted to the sphere are those points where

$$\mathbf{b}(\vec{X}, \hat{\mathbf{s}}_0) \parallel \hat{\mathbf{s}}_0.$$

For the $\vec{X}_0$ case

$$\mathbf{b}(\vec{X}_0, \hat{\mathbf{r}}) = \mathbf{0},$$

so there is a global maximum for the unrestricted function at $\hat{\mathbf{s}}_0 = \hat{\mathbf{r}}$, which must be a unique global maximum on the sphere. Maximality is checked by calculating $\mathbf{A}$. We find:

$$\mathbf{A}(\vec{X}_0, \hat{\mathbf{r}}) = \sum_{i=1}^{m} (1/m) \frac{\hat{\mathbf{a}}_i \hat{\mathbf{a}}_i}{1 - (\hat{\mathbf{a}}_i \cdot \hat{\mathbf{r}})^2}. \qquad (5.66)$$

This is a positive definite symmetric matrix so the Hessian is negative definite and the point is a maximum.[9]

More generally the solution to the stationarity condition depends on $\vec{X}$ in a complicated way.

Ignoring this, assume that $\hat{\mathbf{s}}_0$ is known. Then about this maximum the unrestricted function is given locally by:

$$\begin{aligned} F(\vec{X}, \mathbf{s}) &= F(\vec{X}, \hat{\mathbf{s}}_0) \\ &+ \mathbf{b} \cdot (\mathbf{s} - \hat{\mathbf{s}}_0) \\ &- \frac{1}{2}(\mathbf{s} - \hat{\mathbf{s}}_0) \mathbf{A} (\mathbf{s} - \hat{\mathbf{s}}_0), \qquad (5.67) \end{aligned}$$

where $\mathbf{b}$ and $\mathbf{A}$ are evaluated at $(\vec{X}, \mathbf{s}_0)$. From this series we wish to extract the local behaviour on the sphere.

Choose the z-axis along $\mathbf{s}_0$ and write

$$\hat{\mathbf{s}} = \left(x, y, \sqrt{\{1 - x^2 - y^2\}} \right)$$

---

[9]It is strictly positive because of the stipulation that the set A_m should span $\mathbf{R}^3$.

then approximately,

$$(\hat{s} - \hat{s}_0) = \left(x, y, -1/2(x^2 + y^2)\right).$$

It is helpful to introduce projection matrices,

$$\mathbf{P}_\perp = I - \hat{s}_0\hat{s}_0$$

and

$$\mathbf{P}_\| = \hat{s}_0\hat{s}_0.$$

Then the term in **b** is

$$\mathbf{b} \cdot (\hat{s} - \hat{s}_0) = \mathbf{b}\mathbf{P}_\|(x, y, 0)^T - \frac{1}{2}\mathbf{b}\mathbf{P}_\perp(0, 0, x^2 + y^2)^T.$$

However, $\mathbf{b}\mathbf{P}_\| = \mathbf{0}$; so setting $b_z = \mathbf{b} \cdot \hat{s}_0$ gives

$$\mathbf{b} \cdot (\hat{s} - \hat{s}_0) \simeq -\frac{1}{2}b_z\left(x^2 + y^2\right). \tag{5.68}$$

This is a second order term arising from inserting a local paraboloidal approximation to the sphere surface into the first–order part of the Taylor series. Here we are composing two power series which is perfectly legitimate.

Doing the same for the quadratic term, we need only retain terms of second order. One finds that,

$$(\hat{s} - \hat{s}_0)\mathbf{A}(\hat{s} - \hat{s}_0) \simeq (x, y)\,\mathbf{P}_\perp\mathbf{A}\mathbf{P}_\perp\,(x, y)^T. \tag{5.69}$$

The notation deserves explanation. We mean that part of matrix **A** that acts in the plane perpendicular to **b**. This can be thought of as a 2×2 submatrix being the restriction of **A** to the tangent plane. The inclusion of projection matrices does this for us, it also aids explication of the matrix elements. Let,

$$\mathbf{a}_{\perp i} = \mathbf{P}_\perp\hat{\mathbf{a}}_i,$$

note that this has length,

$$\|\,\mathbf{a}_{\perp i}\,\| = \sqrt{\{1 - (\hat{\mathbf{a}}_i \cdot s_0)^2\}}$$

and then define,

$$\mathbf{A}_\perp(\vec{X}, \hat{s}_0) \equiv \sum_{i=1}^{m} \frac{\mathbf{a}_{\perp i}\mathbf{a}_{\perp i}}{m} \left\{ \frac{2(1 - 2X_i\hat{\mathbf{a}}_i \cdot \hat{s}_0 + (\hat{\mathbf{a}}_i \cdot \hat{s}_0)^2)}{(1 - (\hat{\mathbf{a}}_i \cdot \hat{s}_0)^2)^2} \right\}. \tag{5.70}$$

Observe from eq.(5.66) and the length relation noted above that in the $\vec{X}_0$ case this assumes the simple form,

$$\mathbf{A}_\perp(\vec{X}_0, \hat{\mathbf{r}}) = \sum_{i=1}^{m}(1/m)\hat{\mathbf{a}}_{\perp i}\hat{\mathbf{a}}_{\perp i}. \tag{5.71}$$

Here only normalised projected vectors are involved. We shall deal with this matrix in detail. However, for the moment its more general cousin is retained.

Action along b has been discarded so we are left with a 2×2 matrix acting on the tangent plane, where we are free to rotate the coordinates (x, y). The local approximation reduces to an expression solely in terms of these. It is

$$F(\vec{X}, \hat{\mathbf{s}}) = F(\vec{X}, \hat{\mathbf{s}}_0) - \frac{1}{2}(x, y)\left[\mathbf{A}_\perp + b_z\mathbf{I}\right](x, y)^T. \tag{5.72}$$

Here the quadratic form is negative definite provided $b_z > -\max\{\lambda_1, \lambda_2\}$, where these are the eigenvalues of $\mathbf{A}_\perp$. In the $\vec{X}_0$ case $b_z = 0$ so the curvature correction disappears. We are interested in behaviour local to this condition so no problem is anticipated there.

Substitution of the above equation into eq.(5.63) yields the approximation,

$$
\begin{aligned}
\mathcal{N}(\vec{X}) &\simeq \exp N\left\{F(\vec{X}, \hat{\mathbf{s}}_0)\right\} \int_{-\infty}^{+\infty}\int_{-\infty}^{+\infty} \exp -\frac{N}{2}\left\{\lambda_1^* x^2 + \lambda_2^* y^2\right\} dx\, dy \\
&= \frac{2\pi}{N}\left\{\det[\mathbf{A}_\perp + b_z\mathbf{I}]\right\}^{-1/2} \times \exp N\left\{F(\vec{X}, \hat{\mathbf{s}}_0)\right\}.
\end{aligned}
$$

In this expression $\lambda_i^* = \lambda_i + b_z$, $i = 1, 2$ are the corrected eigenvalues of $\mathbf{A}_\perp$, which are unchanged for the $\vec{X}_0$ case.

From eq.(5.61) it is clear that dividing the above expression by 4π, so as to obtain a normalised sphere integration in eq.(5.63), converts this to the prior probability of outcome $\vec{X}$. That is,

$$p(\vec{X}) \simeq \frac{1}{2N}\left\{\det[\mathbf{A}_\perp + b_z\mathbf{I}]\right\}^{-1/2} \times \exp N\left\{F(\vec{X}, \hat{\mathbf{s}}_0)\right\}. \tag{5.73}$$

Looking at eq.(5.62) we obtain an approximate expression for $G(\vec{X}, \hat{\mathbf{r}})$ as,

$$
\begin{aligned}
G(\vec{X}, \hat{\mathbf{r}}) &\simeq 2N\left\{\det[\mathbf{A}_\perp + b_z\mathbf{I}]\right\}^{+1/2} \\
&\quad \times \exp -\frac{N}{2}\left\{(\hat{\mathbf{r}} - \hat{\mathbf{s}}_0)[\mathbf{A}_\perp + b_z\mathbf{I}](\hat{\mathbf{r}} - \hat{\mathbf{s}}_0)\right\}. \tag{5.74}
\end{aligned}
$$

Recall that this is the inferred distribution for the true state. Hence, eq.(5.73) tells how probable a given outcome is over all possible inputs, whilst eq.(5.74) is the actual output given in terms of the observed data and the implicit maximum $\hat{s}_0[\vec{X}]$.

We now pause to discuss some of the more immediate conclusions arising from these expressions.

5.7.3   Observations concerning G

The anisotropy matrix $\mathbf{A}$ is independent of N. Therefore, it is clear that confidence increases for large N. The outcome $\vec{X}$ could be considered continuous in which case each value determines a particular stationary point. For large N further observations will not move the point $\vec{X}$ very much, however the inferred distribution continues to become more peaked about that direction $\hat{s}_0$ reported by the apparatus as corresponding to the state being measured. Thus it is clear that perfect knowledge of a preparation is possible provided an inexhaustible supply of systems are available.

Another way to think about this is that the increasing density of $\vec{X}$ in its m–cube will place an ever finer net upon the surface of the sphere. Whenever the apparatus is used to observe a particular preparation the inferred state is at each step assigned a single cell of this net containing contracting, ever multiplying, cells. The probability of a wrong assignment is of vanishing measure as $N \to \infty$ in just the same way as Bernoulli convergence of probability[33, p.117].

There is of course an infinite amount of information associated with an infinitesimally fine net. We make this point to emphasise that the entropy associated with quantum states,

$$S = -\mathrm{Tr}\left[\rho \log \rho\right],$$

really has very little to do with how well the state is known. It has more to do with measuring the degree to which the state is mixed. In this formalism, assigning a particular pure state to an a priori pure ensemble involves infinite acquisition of Correlation Information, whereas $S \in [0, \log d]$, of necessity. The extra knowledge concerns the precise direction of $\hat{r}$. Nevertheless, it is possible to draw a link between the two through calculation of the asymptotic inferred density matrix, which we do later.

We have yet to calculate $g(\hat{\mathbf{r}})$, this can be done by working with eq.(5.62) as an integral representation of $G(\vec{X}, \hat{\mathbf{r}})$ and then developing from this an expression for the second partial derivative of its logarithm. Before embarking upon this we shall compare what has been learned so far with numerical experiments. We also elaborate upon the nature of the $\vec{X}_0$ condition and so arrive at a procedure for inverting data for general sets A_N in terms of a matrix similar to $A_\perp$.

5.7.4 Numerical tests of G

In order to fix ideas we compare the analytic expression for asymptotic G with those derived from numerical experiments with a repeated orthogonal triplet for $N = 60$.

Of interest are: the behaviour of the weight function $w(\vec{X})$ defined in eq.(5.50), and that of the prior probabilities given by eq.(5.73).

For the weight function we simply calculate the product of the binomial factors as a function of the parameter

$$R = \sqrt{\{X_1^2 + X_2^2 + X_3^3\}}.$$

It is expected that this should concentrate about unity. This is obvious if one observes that the $\vec{X}_0$ condition becomes:

$$\vec{X}_0 = (\hat{\mathbf{r}} \cdot \hat{\mathbf{x}}, \hat{\mathbf{r}} \cdot \hat{\mathbf{y}}, \hat{\mathbf{r}} \cdot \hat{\mathbf{z}}),$$

for some $\hat{\mathbf{r}}$.

The result of this calculation is shown in figure(5.4). There is a clear peaking which is underemphasised because this is a plot of individual normalised weight values. More properly we should plot an histogram on bins of R. However, this obscures the interesting heirarchy of normal curves. Recall $\vec{X}$ occupys a finite lattice. In this case it has spacing $2/n = 1/10$. The behaviour can then be understood as resulting from a succession of lattice shells in the 3–cube which are approximations to the spherical shell realised in the limit of $n \to \infty$.

Such behaviour is standard law–of–large–numbers material. Pursuing this line for an unrestricted set A_N, let us define a reduced outcome vector:

$$\vec{S}[\Phi_N] \equiv \sum_{i=1}^{N} (1/N)\sigma_i \hat{\mathbf{a}}_i,$$

81

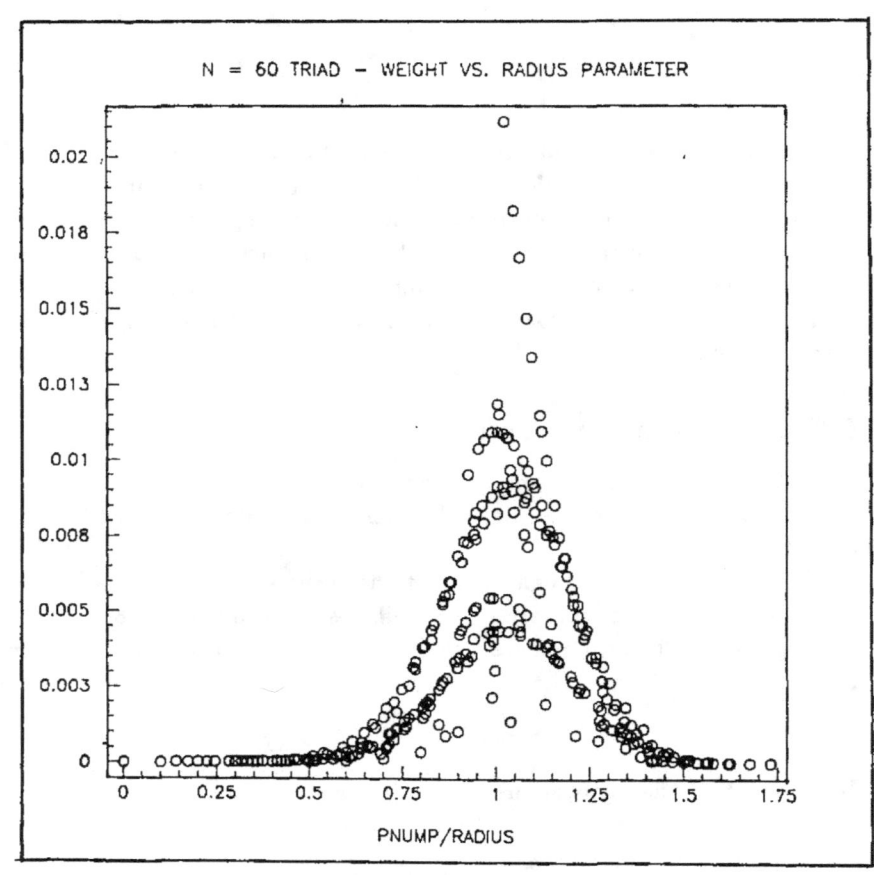

Figure 5.4: Plot of weight values, $w(R)$, versus the radius parameter, R. This calculation is for the $N = 60$ orthogonal triplet. A family of normal–like curves appears superposed.

and the matrix:

$$\mathbf{M} \equiv \sum_{i=1}^{N} (1/N) \hat{\mathbf{a}}_i \hat{\mathbf{a}}_i.$$

Then, in the limit, the true $\hat{\mathbf{r}}$ is given by

$$\lim_{N \to \infty} \langle \vec{S} \rangle_{\Phi_N} = \mathbf{M} \hat{\mathbf{r}}.$$

Here use has been made of the fact that $\langle \sigma_i \rangle = \hat{\mathbf{r}} \cdot \hat{\mathbf{a}}_i$ if one thinks of $\hat{\mathbf{r}}$ as fixed and a sum is taken over all outcomes with probability

$$p(\sigma_i) = \frac{1}{2}(1 + \sigma_i \hat{\mathbf{a}}_i \cdot \hat{\mathbf{r}}).$$

This is a more general statement of the $\vec{X}_0$ condition. Obviously, forming the corresponding $\vec{S} \in \mathbf{R}^3$:

$$\vec{S}[\vec{X}] = \sum_{i=1}^{m} (1/m) X_i \hat{\mathbf{a}}_i,$$

from $\vec{X} \in \mathbf{R}^m$ and applying $\mathbf{M}^{-1}$ yields

$$\hat{\mathbf{r}} = \mathbf{M}^{-1} \vec{S}[\vec{X}]$$

as a necessary result of the $\vec{X}_0$ condition. We should interpret this as a way of inverting the observed data to find the stationary point $\hat{\mathbf{s}}_0$ appearing in eq.(5.74).

Two features must be noted: for $m > 3$ the converse of the above statement is no longer true. The map is no longer one–one from $\vec{X}$ to $\vec{S}$ and so it is possible to obtain a $\hat{\mathbf{r}}$ via this route which does not reproduce $\vec{X}$ upon taking its inner product with the $\hat{\mathbf{a}}_i$. However, this is not a problem because the weight function guarantees that there is vanishing measure for observing such $\vec{X}$ in an experimant upon a pure preparation. If the state observed is indeed pure then only good results will be seen for large N.[10]

---

[10] We must therefore consider what happens if the a priori assumption of a pure state is relaxed. In fact an experiment will tell us if this is wrong by generating a refutation, at large N, in the form $(\| \mathbf{M}^{-1} \vec{S} \| - 1)^2 > \alpha/N$, where α is some confidence parameter. The unused parts of the m-cube of $\vec{X}$-space tell about impure states. They were thrown away by assuming $\| \mathbf{r} \| = 1$ in eq.(5.53). The generalisation to impure inputs is obvious (integration measure of sphere plus interior). We do not follow it further because $A_\perp$ loses its nice form.

A compact statement of this conclusion is that we expect $\hat{\mathbf{s}}_0$ to be close to $\mathbf{M}^{-1}\vec{S}[\vec{X}]$ for any $\vec{X}$ arising from a real experiment. The matrix M is A_m–dependent and of unit trace. For Platonic solid sets it is always $1/3\mathbf{I}$, so data from these measurements is particularly easy to invert.

In figure(5.5) the logarithm of exact prior probabilities is plotted against the radius parameter, revealing linear behaviour. Looking at eq.(5.73) it is clear that the linearity of $F(\vec{X}, \hat{\mathbf{s}}_0)$ in X_i is largely responsible for this. Note that the highest probability outcomes lie at suppressed values of the radius parameter. They have individually high probability but there are not many of them. The result at radius paramater $R = 0$ has lowest probability. It corresponds to ambivalent outcomes(equal yes–no results) upon each basis direction. Such an outcome would suggest an unpolarised input state and does give this density matrix as an output, if the a priori pure state condition is relaxed.

Finally we explore the accuracy of the asymptotic expression eq.(5.73) for two cases. In both instances the true stationary point is located by a simulated annealing search to maximise $p(\vec{X}|\hat{\mathbf{r}})$. Then this is used to calculate $G(\vec{X}, \hat{\mathbf{r}})$. Comparison is made with exact probabilities from the triplet formula. In figure(5.6) the curvature correction b_z is set to zero, whilst it is included for figure(5.7).

There is a clear improvement for $R > 1$ when the correction is included. This can be understood by recognising that in this regime the surface curvature emphasises the peak at the stationary point and thereby improves the quality of approximation. The opposite is true for interior points and we can observe no great improvement for such points. At $R = 1$ there is no change because b_z is already zero.

From now on it is not necessary to include the curvature correction as no integrations are to be done which are not local to $b_z = 0$. We are left with:

$$G(\vec{X}, \hat{\mathbf{r}}) \simeq 2N\{\det \mathbf{A}_\perp\}^{+1/2}$$
$$\times \exp -\frac{N}{2}\{(\hat{\mathbf{r}} - \hat{\mathbf{s}}_0)\mathbf{A}_\perp(\hat{\mathbf{r}} - \hat{\mathbf{s}}_0)\}. \qquad (5.75)$$

From which it is simple to calculate:

$$\log G(\vec{X}_0, \hat{\mathbf{r}}) = \log N + 1/2\log\{4\det \mathbf{A}_\perp(\hat{\mathbf{r}})\}. \qquad (5.76)$$

This verifies the assertion made about eq.(5.60) upon page 74. The optimality of A_m is measured by its associated anisotropy matrix $\mathbf{A}_\perp(\hat{\mathbf{r}})$ where it is only

84

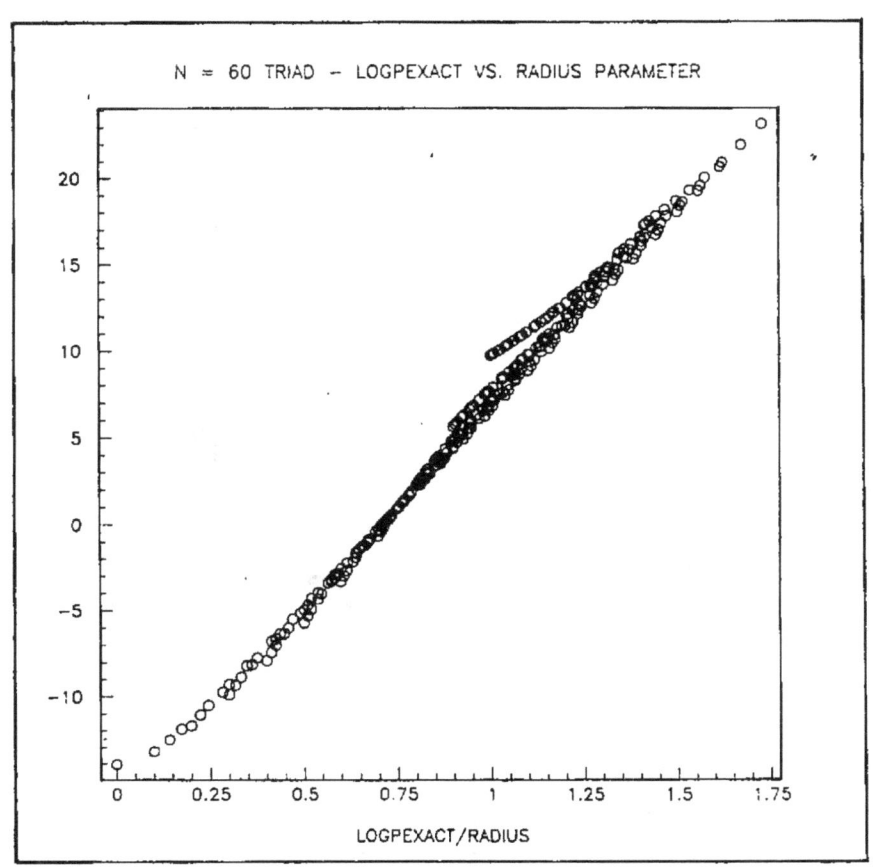

Figure 5.5: Plot of log prior probabilities against the radius parameter, R. This calculation is for the $N = 60$ orthogonal triplet from the exact expression. A straight line trend is indicated with some fine structure as observed for the corresponding plot of weights.

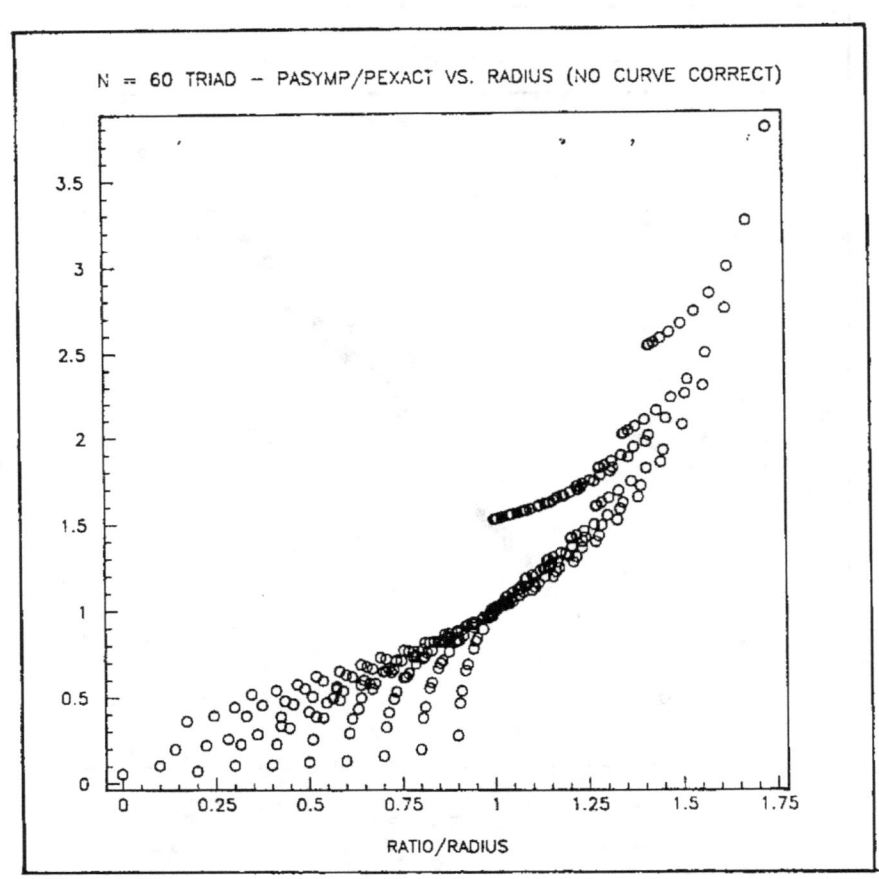

N = 60 TRIAD — PASYMP/PEXACT VS. RADIUS (NO CURVE CORRECT)

Figure 5.6: Steepest descent excluding correction for sphere curvature.

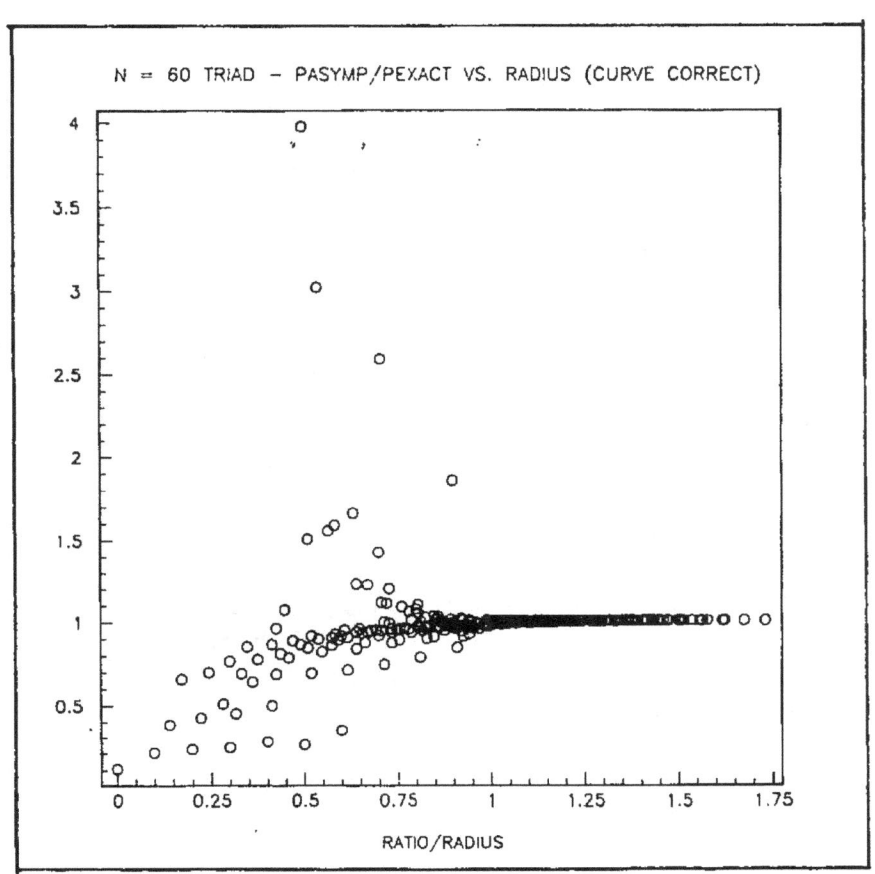

Figure 5.7: Steepest descent including correction for sphere curvature.

necessary to deal with those of the special form given in eq.(5.71). There is still the matter of calculating $g(\hat{\mathbf{r}})$, which we deal with now.

5.7.5 Calculation of g

Recall the definitions:

$$g(\hat{\mathbf{r}}) \equiv \frac{1}{2n} \sum_{i=1}^{m} \left(1 - (\hat{\mathbf{a}}_i \cdot \hat{\mathbf{r}})^2\right) \frac{\partial^2}{\partial X_i^2} \left[\log G(\vec{X}, \hat{\mathbf{r}})\right]_{\vec{X}_0}$$

and

$$G(\vec{X}, \hat{\mathbf{r}}) \equiv \frac{1}{\mathcal{N}(\vec{X})} \exp N\left\{F(\vec{X}, \hat{\mathbf{r}})\right\}.$$

We shall need the first partial derivative of $F(\vec{X}, \hat{\mathbf{r}})$, which is

$$\frac{\partial}{\partial X_i} F(\vec{X}, \hat{\mathbf{r}}) = \frac{n}{2} \log \left[\frac{1 + \hat{\mathbf{a}}_i \cdot \hat{\mathbf{r}}}{1 - \hat{\mathbf{a}}_i \cdot \hat{\mathbf{r}}}\right].$$

However, rather than use this, we approximate it by a Taylor series about the maximum of G. In terms of $\hat{\mathbf{s}}$ we set

$$\frac{\partial}{\partial X_i} F(\vec{X}, \hat{\mathbf{s}}) \simeq \frac{n}{2} \log \left[\frac{1 + \hat{\mathbf{a}}_i \cdot \hat{\mathbf{s}}_0}{1 - \hat{\mathbf{a}}_i \cdot \hat{\mathbf{s}}_0}\right]$$
$$+ \frac{n}{1 - (\hat{\mathbf{a}}_i \cdot \hat{\mathbf{s}}_0)^2} \times \hat{\mathbf{a}}_i \cdot (\hat{\mathbf{s}} - \hat{\mathbf{s}}_0)$$
$$+ \frac{n \, \hat{\mathbf{a}}_i \cdot \hat{\mathbf{s}}_0}{(1 - (\hat{\mathbf{a}}_i \cdot \hat{\mathbf{s}}_0)^2)} \times (\hat{\mathbf{a}}_i \cdot (\hat{\mathbf{s}} - \hat{\mathbf{s}}_0))^2. \qquad (5.77)$$

For compactness, we label $1/n$ times these terms as: c_0, c_1 and c_2. The subscript indicates the degree in $\hat{\mathbf{s}}$. After these preliminaries we are ready to calculate g.

Taking second partial derivatives of $\log G$ it is clear that the numerator makes no contribution (F is linear in X_i). Therefore,

$$\frac{\partial^2}{\partial X_i^2} \left[\log G(\vec{X}, \hat{\mathbf{r}})\right]_{\vec{X}_0} = \frac{\partial^2}{\partial X_i^2} \left[-\log \mathcal{N}(\vec{X})\right]_{\vec{X}_0}, \qquad (5.78)$$

which can be written as,

$$\frac{\partial^2}{\partial X_i^2} \left[\log G(\vec{X}, \hat{\mathbf{r}})\right]_{\vec{X}_0} = \frac{1}{\mathcal{N}^2} \left[\mathcal{N} \frac{\partial^2 \mathcal{N}}{\partial X_i^2} - \frac{\partial \mathcal{N}}{\partial X_i} \frac{\partial \mathcal{N}}{\partial X_i}\right]_{\vec{X}_0}. \qquad (5.79)$$

88

This reduces the quantity to one solely in terms of integrations at $\vec{X}_0$. It is necessary to calculate:

$$\frac{\partial \mathcal{N}}{\partial X_i}[\vec{X}_0] = \int \frac{n}{2} \log \left[\frac{1 + \hat{a}_i \cdot \hat{s}}{1 - \hat{a}_i \cdot \hat{s}} \right] \exp N\left\{ F(\vec{X}_0, \hat{s}) \right\} d\Omega_{\hat{s}}, \quad (5.80)$$

$$\frac{\partial^2 \mathcal{N}}{\partial X_i^2}[\vec{X}_0] = \int \frac{n}{2} \log^2 \left[\frac{1 + \hat{a}_i \cdot \hat{s}}{1 - \hat{a}_i \cdot \hat{s}} \right] \exp N\left\{ F(\vec{X}_0, \hat{s}) \right\} d\Omega_{\hat{s}}. \quad (5.81)$$

There is no need to approximate F in these integrals yet. However, we do approximate the log prefactors. For further compactnes we define,

$$\langle f(\hat{s}) \rangle \equiv \int f(\hat{s}) \exp N\left\{ F(\vec{X}_0, \hat{s}) \right\} d\Omega_{\hat{s}}.$$

Then, eq.(5.80) is

$$I_1 = n \times \langle c_0 + c_1 + c_2 \rangle,$$

whilst eq.(5.81) becomes

$$I_2 = n^2 \times \langle (c_0 + c_1 + c_2)^2 \rangle.$$

In the same notation we need,

$$\frac{\partial^2}{\partial X_i^2} \left[\log G(\vec{X}, \hat{r}) \right]_{\vec{X}_0} = \frac{\langle 1 \rangle I_2 - I_1^2}{\langle 1 \rangle \langle 1 \rangle}. \quad (5.82)$$

Making use of $\langle c_0 \rangle = c_0 \langle 1 \rangle$ and retaining only terms of second degree or less in $\hat{s}$ gives,

$$\langle 1 \rangle I_2 - I_1^2 = n^2 \langle c_0 \rangle \langle c_0 + 2(c_1 + c_2) + c_0^{-1} c_1^2 \rangle$$
$$- n^2 \langle c_0 \rangle \langle c_0 + 2(c_1 + c_2) \rangle$$
$$- n^2 \langle c_1 \rangle \langle c_1 \rangle.$$

Making the obvious cancellations leaves,

$$\frac{\partial^2}{\partial X_i^2} \left[\log G(\vec{X}, \hat{r}) \right]_{\vec{X}_0} = n^2 \frac{\langle 1 \rangle \langle c_1^2 \rangle - \langle c_1 \rangle \langle c_1 \rangle}{\langle 1 \rangle \langle 1 \rangle}.$$

So far, F has remained exact. We now include the approximations made previously. With $\hat{s}$ the intended integration variable and $\hat{s}_0 = \hat{r}$ as the $\vec{X}_0-$determined stationary point, they are:

$$(\hat{s} - \hat{r}) = \left(x, y, -1/2(x^2 + y^2) \right), \quad (5.83)$$

$$F(\vec{X}, \hat{s}) = F(\vec{X}, \hat{r}) - \frac{1}{2}(x, y) \mathbf{A}_\perp (x, y)^T. \quad (5.84)$$

89

The exponential factors in $F(\vec{X}, \hat{\mathbf{r}})$ cancel between numerator and denominator. If the appropriate substitutions are made then one finds that,

$$O(\langle c_1 \rangle) = O(\langle c_1^2 \rangle) = O(1/N).$$

However, the first term appears squared so we can discard it. This yields,

$$\frac{\partial^2}{\partial X_i^2} \left[\log G(\vec{X}, \hat{\mathbf{r}}) \right]_{\vec{X}_0} = n^2 \langle c_1^2 \rangle / \langle 1 \rangle.$$

Now we form $g(\hat{\mathbf{r}})$

$$g(\hat{\mathbf{r}}) = \frac{1}{2n} \sum_{i=1}^{m} \left(1 - (\hat{\mathbf{a}}_i \cdot \hat{\mathbf{r}})^2 \right) \times n^2 \langle c_1^2 \rangle / \langle 1 \rangle$$

All of the new terms can be taken inside the expectation. Upon substitution of

$$c_1 = \frac{n}{1 - (\hat{\mathbf{a}}_i \cdot \hat{\mathbf{r}})^2} \times \hat{\mathbf{a}}_i \cdot (\hat{\mathbf{s}} - \hat{\mathbf{r}}),$$

some terms cancel and we find that,

$$\sum_{i=1}^{m} \left(1 - (\hat{\mathbf{a}}_i \cdot \hat{\mathbf{r}})^2 \right) c_1^2 = m \times (\hat{\mathbf{s}} - \hat{\mathbf{r}}) \mathbf{A}_\perp (\hat{\mathbf{s}} - \hat{\mathbf{r}}).$$

Therefore,

$$g(\hat{\mathbf{r}}) = \frac{N}{2} \times \langle (\hat{\mathbf{s}} - \hat{\mathbf{r}}) \mathbf{A}_\perp (\hat{\mathbf{s}} - \hat{\mathbf{r}}) \rangle / \langle 1 \rangle,$$

where $N = mn$ has been substituted. The expectation can now be calculated by using the result,

$$\frac{1}{2} \langle (\hat{\mathbf{s}} - \hat{\mathbf{r}}) \mathbf{A}_\perp (\hat{\mathbf{s}} - \hat{\mathbf{r}}) \rangle = -\frac{d}{dN} \langle 1 \rangle. \qquad (5.85)$$

Recall that,

$$\langle 1 \rangle = \frac{2\pi}{N} \{ \det \mathbf{A}_\perp \}^{-1/2},$$

whence,

$$\frac{d}{dN} \langle 1 \rangle = -1/N \times \langle 1 \rangle.$$

Substitution of this into the expression for $g(\hat{\mathbf{r}})$ gives the final result,

$$g(\hat{\mathbf{r}}) = -1,$$

as required.

90

5.7.6 Asymptotic formula

It follows that the complete asymptotic expression corresponding to substitution of eq.(5.76) and g into eq.(5.60) is,

$$
\begin{aligned}
\{\hat{\mathbf{r}}, \Phi_N\} &= \log N - 1 \\
&\quad + 1/2 \int \log \left\{ 4 \det \mathbf{A}_\perp(\hat{\mathbf{r}}) \right\} d\hat{\Omega}_{\hat{\mathbf{r}}}.
\end{aligned}
\tag{5.86}
$$

There is a leading A_m independent information gain that grows as the logarithm of the number of trials. The second term embodies the apparatus dependence and is independent of the number of trials. Defining,

$$
\beta[A_m] \equiv -1/2 \int \log \left\{ 4 \det \mathbf{A}_\perp(\hat{\mathbf{r}}) \right\} d\hat{\Omega}_{\hat{\mathbf{r}}},
$$

it is clear that differing β have the effect of rescaling N. Shortly it will become clear that $\beta \geq 0$. In terms of β, we have,

$$
\{\hat{\mathbf{r}}, \Phi_N\} = \log e^{-\beta} N - 1.
$$

It follows that differing geometries A_m and $A'_{m'}$ acquire equal information after numbers of trials N and N' that are in the ratio $e^{-\beta} N : e^{-\beta'} N'$. Therefore, in terms of N we require $N' = e^{(\beta'-\beta)} N$ observations to have the same information. This is a constant multiplier that exists for any two rival geometries. Clearly minimising β maximises the Correlation Information.

Turning now to consideration of $\mathbf{A}_\perp$, recall the definition given on page 79,

$$
\mathbf{A}_\perp(\hat{\mathbf{r}}) = \sum_{i=1}^{m} (1/m)\hat{\mathbf{a}}_{\perp i}\hat{\mathbf{a}}_{\perp i},
$$

where the $\hat{\mathbf{r}}$ dependence comes from,

$$
\mathbf{a}_{\perp i} = \hat{\mathbf{a}}_i - (\hat{\mathbf{a}}_i \cdot \hat{\mathbf{r}})\hat{\mathbf{r}}.
$$

Notice that this matrix remains well defined for arbitrary sets A_N so we expect that the Correlation Information given above applies in general. Also in virtue of the normalisation of projected $\mathbf{a}_i$, it follows that,

$$
\mathrm{Tr}\mathbf{A}_\perp(\hat{\mathbf{r}}) = 1, \qquad \forall\, \hat{\mathbf{r}}.
$$

91

Furthermore, since we demanded that the A_m set should span $\mathbf{R}^3$, it is clear that neither of the two eigenvalues is ever zero. Since the matrix is positive definite it follows from the unit trace condition that,

$$0 < 4 \det \mathbf{A}_\perp(\hat{\mathbf{r}}) \leq 1$$

and so $\beta \geq 0$ as suggested above. Equality is gained only when both eigenvalues are equal to one half.

Given this condition it is suggested that,

$$\{\hat{\mathbf{r}}, \Phi_N\} \leq \log N - 1, \tag{5.87}$$

with equality if and only if

$$4 \det \mathbf{A}_\perp(\hat{\mathbf{r}}) = 1, \qquad \forall\, \hat{\mathbf{r}}.$$

We shall soon discover that this is only possible for those A_N which approximate a uniform distribution of $\hat{\mathbf{a}}_i$ over the sphere for large N.

A simple intuitive grasp of the nature of the anisotropy matrix is provided by recognising that, if one looks along the direction of $\hat{\mathbf{r}}$, then the set A_m is projected onto a unit disk in the plane tangent to $\hat{\mathbf{r}}$. The $\hat{\mathbf{a}}_{\perp i}$ populate the rim of this disk and move around it as the orientation of $\hat{\mathbf{r}}$ is changed relative to A_m. Of course $\mathbf{A}_\perp$ is invariant under inversions of the $\hat{\mathbf{a}}_i$ as expected. This translates to inversion symmetry on the disk.

Thinking this way makes the above non-singular assertion obvious, but we prove it anyway. Also it shows that the upper bound to $\det \mathbf{A}$ is clearly attained only for those $\hat{\mathbf{r}}$ where A_m appears isotropic on the projection disk.

The observation of unit trace makes calculation of the determinant easy. Consider the matrix elements of this 2×2 matrix for a arbitrary basis $(\hat{\mathbf{x}}, \hat{\mathbf{y}})$ on the disk. The determinant is

$$\det \mathbf{A}_\perp = (1/m)^2 \begin{vmatrix} x_i x_i & x_i y_i \\ y_i x_i & y_i y_i \end{vmatrix}, \tag{5.88}$$

where summation over $i = 1, m$ is implicit. Defining the angle between two vectors on the disk as α_{ij}, it is not difficult to show that,

$$\det \mathbf{A}_\perp = (1/m)^2 \sum_{i<j} \sin^2 \alpha_{ij}. \tag{5.89}$$

92

In the special case $m = 3$, there is a simple formula in terms of the three basis directions labelled, $\hat{\mathbf{a}}, \hat{\mathbf{b}}, \hat{\mathbf{c}}$. It is

$$\det \mathbf{A}_\perp(\hat{\mathbf{r}}) = (1/9)\left[(\hat{\mathbf{a}} \cdot \hat{\mathbf{b}} \wedge \hat{\mathbf{c}})^2(1 - (\hat{\mathbf{a}} \cdot \hat{\mathbf{r}})^2)(1 - (\hat{\mathbf{b}} \cdot \hat{\mathbf{r}})^2)(1 - (\hat{\mathbf{c}} \cdot \hat{\mathbf{r}})^2)\right]^{-1}$$
$$\times \left\{ (\hat{\mathbf{r}} \cdot \hat{\mathbf{b}} \wedge \hat{\mathbf{c}})^2(1 - (\hat{\mathbf{a}} \cdot \hat{\mathbf{r}})^2) + (\hat{\mathbf{r}} \cdot \hat{\mathbf{a}} \wedge \hat{\mathbf{c}})^2(1 - (\hat{\mathbf{b}} \cdot \hat{\mathbf{r}})^2) \right.$$
$$\left. + (\hat{\mathbf{r}} \cdot \hat{\mathbf{a}} \wedge \hat{\mathbf{b}})^2(1 - (\hat{\mathbf{c}} \cdot \hat{\mathbf{r}})^2) \right\}$$

Notice that general formula has zeros if and only if,

$$\exists\, \hat{\mathbf{r}} \vdash \quad \alpha_{ij} = 0 \quad \forall\, (i,j).$$

So we have verified that $\mathbf{A}_\perp$ is non-singular unless A_m is a coplanar or collinear set.

Given these formulæ it is possible to calculate $\mathbf{A}_\perp$ at arbitrary $\hat{\mathbf{r}}$ and so numerically integrate the term $\log\{4\det \mathbf{A}_\perp(\hat{\mathbf{r}})\}$ over the sphere. One minor point of interest is the question of what happens to the determinant near a basis direction.

Concerning the behaviour of $\mathbf{A}_\perp$ in the neighbourhood of the vector $\hat{\mathbf{a}}_j$, it is helpful to consider the matrix,

$$\mathbf{A}_\perp^* \equiv \sum_{i \neq j} 1/(m-1)\hat{\mathbf{a}}_{\perp i}\hat{\mathbf{a}}_{\perp i}.$$

If this is $1/2\mathbf{I}$, then the local behaviour in a small disk about $\hat{\mathbf{r}} = \hat{\mathbf{a}}_j$ is continuous. In general this is not the case, but the problem is not serious as it only happens at isolated points in the region over which the integration is to take place.

The Platonic solid sets are nice in that they do satisfy the above identity condition on basis directions. The interesting feature of the uniform sampling set mentioned above is that it does so everywhere.

Wherever the above condition is true, it follows that $A_\perp$ assumes the local form,

$$\mathbf{A}_\perp = (1 - 1/m)\,1/2\mathbf{I} + 1/m\,\hat{\mathbf{a}}_{\perp i}\hat{\mathbf{a}}_{\perp i}.$$

The last term is always a rotated one–dimensional projector, for any direction of approach within the plane tangent to the basis direction.

So we may define,

$$\det \mathbf{A}_\perp(\hat{\mathbf{a}}_i) \equiv (m-1)/m^2 \begin{vmatrix} \frac{1}{2} & 0 \\ 0 & \frac{1}{2} \end{vmatrix} + 1/m^2 \begin{vmatrix} 1 & 0 \\ 0 & 0 \end{vmatrix}.$$

This expression yields,

$$\det \mathbf{A}_\perp(\hat{\mathbf{a}}_i) = 1/4 - 1/m^2.$$

Calculating this determinant at the basis directions for the Platonic solids gives:

Orthogonal triplet A_3 :	2/9,
Tetrahedron	A_4 :	15/64,
Dodecahedron	A_6 :	35/144,
Icosahedron	A_{10} :	99/400.

For the uniform sampling set, the value is given by the same expression at all points of the sphere, because each may be considered to be a basis direction. Taking large m ensures that everywhere upon the sphere,

$$\det \mathbf{A}_\perp(\hat{\mathbf{r}}) = 1/4 - 1/m^2.$$

Hence it is possible to approach the upper bound with such a set.

We shall develop an argument tailored to this special case in the next section. Prior to that, it is useful to develop a picture of the relative merits of the Platonic solid sets. First we note the results of numerical integration to obtain the parameters β. Then we consider contour plots of the determinant upon the sphere. At each $\hat{\mathbf{r}}$ the departure of this from the ideal value of 1/4 measures the relative anisotropy of spatial sampling along the two local eigenvectors in the projection disk at that point.

The numerical results for the Platonic solids are:

Orthogonal triplet β_3 :	0.039 720 7(40),
Tetrahedron	β_4 :	0.023 475 4(09),
Dodecahedron	β_6 :	0.008 829 0(04),
Icosahedron	β_{10} :	0.003 738 0(26).

These were obtained via numerical integration using *NAG* routine *D01FCF*. The bracketed figures indicate the numerical precision as reported by this subroutine. The calculation is single precision.

These are best interpreted in terms of the equation $N' = e^{(\beta'-\beta)}N$. There is little difference. The ratio comparing the icosahedron to the orthogonal triplet is,

$$N_{\text{ico}} = e^{\beta_{10}-\beta_3} N_{\text{tri}} = 0.965 N_{\text{tri}}.$$

Thus approximately 96.5 per cent fewer observations are required for an icosahedral set. Comparing the icosahedron to the ideal uniform measurement set, $\beta = 0$, yields,

$$N_{\text{ico}} = e^{\beta_{10}} N_{\text{ideal}} = 1.0037 N_{\text{ideal}}.$$

Only 0.37 per cent more test subsystems need be available to derive the same information as from an ideal apparatus.

The contour plots shown in figure(5.8) through figure(5.11) display the determinant values upon sections of the sphere. In each case an appropriate symmetry was indentified and the sphere decomposed into a set of spherical tiles. Each tile was scanned for contours in polar fashion so as to minimise distortion. The contours are then projected from the sphere surface onto a plane tangent to the central point.

These figures are more or less self–explanatory. Note that, although the dodecahedron has pentagonal faces, the relevant symmetry is that of its dual figure, namely the icosahedron, and vice–versa. The ideal apparatus would have a uniform value of 1/4 over the entire surface. This is attained along particular directions for the Platonic sets. Such directions could be considered as defining the preference of the apparatus. Observe that basis directions are never optimal and are in fact worst for the orthogonal triplet and for the dodecahedron.

5.8   Uniform Measurement Set

A simple argument is now presented for independent derivation of the Correlation Information for a uniform measurement set. We assume a large number, N, of bases, $\hat{\mathbf{a}}$, distributed uniformly over the sphere. The asymptotic convergence of all projector expectations, plus symmetry considerations, then ensure that there is equal preference for all unit vector outputs. We call these $\hat{\mathbf{s}}$, whilst the input pure preparation remains $\hat{\mathbf{r}}$. The N–trial correlation then becomes,

$$p(\hat{\mathbf{s}}|\hat{\mathbf{r}}) = \frac{1}{\mathcal{N}} \exp N \left\{ \int (1 + \hat{\mathbf{a}} \cdot \hat{\mathbf{s}}) \log \frac{1}{2}(1 + \hat{\mathbf{a}} \cdot \hat{\mathbf{r}}) \, d\hat{\Omega}_{\hat{\mathbf{a}}} \right\}. \qquad (5.90)$$

95

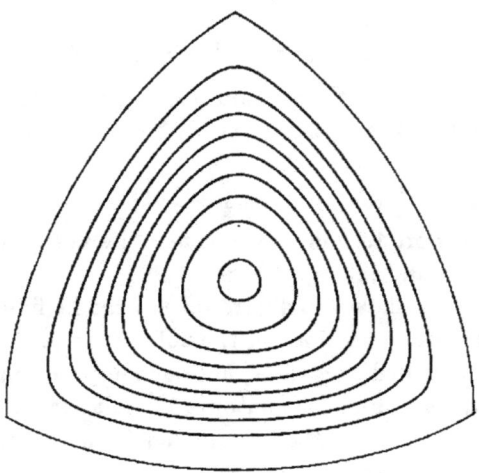

Figure 5.8: Anisotropy contours for the repeated orthogonal triplet. The measurement vectors lie at the three vertices of this projected octant. The inner contour has value 0.2495, decreasing with interval 0.003 towards the edges. The boundary is itself a contour with value 2/9. Greatest sensitivity occurs at the central maximum, where this corresponds to the average of the three basis directions and the best possible value of 1/4 is attained.

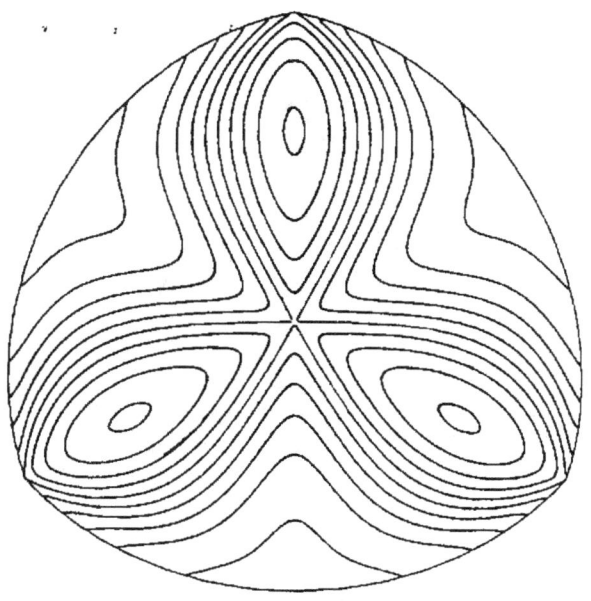

Figure 5.9: Anisotropy contours for the repeated tetrahedron. Three measurement vectors lie at the vertices of this quarter sphere. The fourth lies at the centre. At the basis vectors the anisotropy is 15/64. The value is 0.2495 for the outermost contour decreasing in steps of 0.003. In this case greatest sensitivity, of 1/4, occurs along the directions corresponding to the bisectors of the curved boundary.

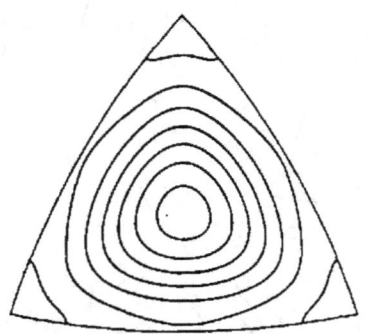

Figure 5.10: One of twenty spherical tiles forming an anisotropy map for the repeated dodecahedron. The three vertices are neighbouring bases. The central point is a maximum, with value 1/4 corresponding to optimal sensitivity and isotropic sampling. On the inner contour the anisotropy is 0.2495 decreasing towards the edge in steps of 0.001. Along a basis direction the anisotropy is 35/144.

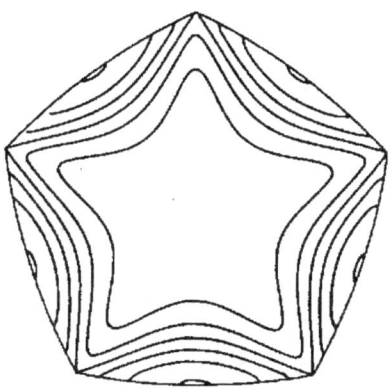

Figure 5.11: One of twelve spherical tiles forming an anisotropy map for the repeated icosahedron. The five vertices are neighbouring bases. The central point is a maximum, with value 1/4 corresponding to optimal sensitivity and isotropic sampling. On the inner contour the anisotropy is 0.2495 decreasing towards the edge in steps of 0.001. Along a basis direction the anisotropy is 99/400.

99

This generalisation follows from consideration of eq.(5.49), where uniform integration over the basis directions replaces the product of individual basis correlations. A factor of one half is lost from the log prefactor because of the two signs present in eq.(5.49). The normalisation must be a constant from symmetry considerations.

Calculating the integral in the exponent we find that,

$$\int (1 + \hat{\mathbf{a}} \cdot \hat{\mathbf{s}}) \log \frac{1}{2}(1 + \hat{\mathbf{a}} \cdot \hat{\mathbf{r}}) \, d\hat{\Omega}_{\hat{\mathbf{a}}} = -1 + \frac{1}{2}\hat{\mathbf{s}} \cdot \hat{\mathbf{r}}. \tag{5.91}$$

For the normalisation, use of the previous result,

$$\int e^{\mathbf{b} \cdot \hat{\mathbf{r}}} d\hat{\Omega}_{\hat{\mathbf{r}}} = |b|^{-1} \sinh |b|,$$

with $|b| = N/2$, gives the result,

$$\mathcal{N} = \int \exp N\{-1 + 1/2\hat{\mathbf{s}} \cdot \hat{\mathbf{r}}\} \, d\hat{\Omega}_{\hat{\mathbf{r}}} = (N/2)^{-1} \sinh(N/2) \times e^{-N}. \tag{5.92}$$

Thus the equivalent of G is,

$$G(\hat{\mathbf{s}}, \hat{\mathbf{r}}) = \frac{N/2}{\sinh N/2} \times e^{N/2\{\hat{\mathbf{s}} \cdot \hat{\mathbf{r}}\}}. \tag{5.93}$$

Approximating $\sinh N/2 \simeq 1/2 e^{N/2}$, for large N, we recover the expected form,

$$G(\hat{\mathbf{s}}, \hat{\mathbf{r}}) = N \times e^{N/2\{\hat{\mathbf{s}} \cdot \hat{\mathbf{r}} - 1\}} \simeq N \times \exp -\frac{N}{2}\{1/2x^2 + 1/2y^2\}, \tag{5.94}$$

where (x, y) are coordinates of the plane tangent to $\hat{\mathbf{s}}$. However, we shall not need this approximation, as the integrals can be done exactly.

Recall that G gives the inferred distribution for $\hat{\mathbf{r}}$, given observed $\hat{\mathbf{s}}$. In terms of the basis outcomes (signs absorbed) this is,

$$\hat{\mathbf{s}} = \mathbf{M}^{-1} \sum_{i=1}^{N} \hat{\mathbf{a}}_i / N.$$

In this case $\mathbf{M} = 1/3\,\mathbf{I}$ and so we need only sum over all of the observed projectors and divide by $3N$.

100

Substituting for the exact G in the Correlation Information, it follows that,

$$\{\hat{\mathbf{s}}, \hat{\mathbf{r}}\} = \int \int \frac{N/2}{\sinh N/2} \times e^{N/2\{\hat{\mathbf{s}} \cdot \hat{\mathbf{r}}\}} \log \left[\frac{N/2}{\sinh N/2} \times e^{N/2\{\hat{\mathbf{s}} \cdot \hat{\mathbf{r}}\}} \right] d\hat{\Omega}_{\hat{\mathbf{s}}} d\hat{\Omega}_{\hat{\mathbf{r}}}. \quad (5.95)$$

This is,

$$\{\hat{\mathbf{s}}, \hat{\mathbf{r}}\} = \log N - \log \left[2 \sinh(N/2) \right] + \frac{N/2}{\sinh N/2} \times N \frac{d}{dN} \left[\frac{\sinh N/2}{N/2} \right]. \quad (5.96)$$

Using,

$$N \frac{d}{dN} \left[\frac{\sinh N/2}{N/2} \right] = - \frac{\sinh N/2}{N/2} + \cosh N/2,$$

we find that,

$$\{\hat{\mathbf{s}}, \hat{\mathbf{r}}\} = \log N - 1 - \log \left[2 \sinh N/2 \right] + N/2 \coth N/2. \quad (5.97)$$

Substituting for sinh and coth this becomes,

$$\{\hat{\mathbf{s}}, \hat{\mathbf{r}}\} = \log N - 1 - \log[e^{N/2} - e^{-N/2}] + N/2 \times \frac{1 + e^{-N}}{1 - e^{-N}}. \quad (5.98)$$

Clearly, this approaches the expected asymptotic form,

$$\{\hat{\mathbf{s}}, \hat{\mathbf{r}}\} \simeq \log N - 1. \quad (5.99)$$

Also, it does so from above, in agreement with the qualitative behaviour identified for triplet measurement in the earlier numerical experiments.

5.9 Upper bound to the Correlation Information

In this section we shall calculate a conjectured upper bound to the Correlation Information. It turns out to be the same as eq.(5.99), showing that the uniform measurement set is indeed optimal.

If use is made of one of the alternative expressions for the Correlation Information listed in §3.4, eq.(3.15) and eq.(3.16), then for a uniform prior, $p_0(\hat{\mathbf{r}}) = 1/4\pi$, we find that,

$$\{\hat{\mathbf{r}}, \Phi_N\} = \sum_{\forall \Phi_N} p(\Phi_N) \times \int p(\hat{\mathbf{r}}|\Phi_N) \log p(\hat{\mathbf{r}}|\Phi_N) \, d\hat{\Omega}_{\hat{\mathbf{r}}}$$

$$- \int \log p_0(\hat{\mathbf{r}}) \, d\hat{\Omega}_{\hat{\mathbf{r}}}. \qquad (5.100)$$

Using the normalisation condition,

$$\int p(\hat{\mathbf{r}}|\Phi_N) \, d\hat{\Omega}_{\hat{\mathbf{r}}} = 1$$

and substituting for $p_0(\hat{\mathbf{r}})$ yields,

$$\{\hat{\mathbf{r}}, \Phi_N\} = \sum_{\forall \Phi_N} p(\Phi_N) \times \int p(\hat{\mathbf{r}}|\Phi_N) \log \left[4\pi p(\hat{\mathbf{r}}|\Phi_N)\right] \, d\hat{\Omega}_{\hat{\mathbf{r}}}. \qquad (5.101)$$

Notice that the integral is a function of Φ_N only, given a certain choice of apparatus, A_N. Therefore, for each A_N, it is certainly true that,

$$\{\hat{\mathbf{r}}, \Phi_N\} \leq \max_{\forall \Phi_N} \left\{ \int p(\hat{\mathbf{r}}|\Phi_N) \log \left[4\pi p(\hat{\mathbf{r}}|\Phi_N)\right] \, d\hat{\Omega}_{\hat{\mathbf{r}}} \right\}. \qquad (5.102)$$

The $p(\hat{\mathbf{r}}|\Phi_N)$ are simply normalised functions on the sphere. Each has the form,

$$\frac{1}{\mathcal{N}} \prod_{i=1}^{N} \frac{1}{2}(1 + \sigma_i \hat{\mathbf{a}}_i \cdot \hat{\mathbf{r}}),$$

with $\mathcal{N}$ the appropriate normalisation. Each integral measures the information of this function. Intuitively, the integral is largest for those functions of the above form which are peakiest. Therefore, we claim that an upper bound to the Correlation Information can be obtained by substituting the most strongly peaked $p(\hat{\mathbf{r}}|\Phi_N)$ that is possible after N trials. We further claim that this function is,

$$p(\hat{\mathbf{r}}|\Phi_N) = \frac{(N+1)}{4\pi} \left(\frac{1 + \hat{\mathbf{a}} \cdot \hat{\mathbf{r}}}{2}\right)^N,$$

where the choice of $\hat{\mathbf{a}}$ does not matter. Such an inferred output corresponds to the all yes outcome for a singlet measurement. As an outcome it has small probability of occurence since it indicates that the true state is very close to the direction that was chosen for the singlet basis. The chances of such a fortutitous choice are low. Nevertheless, were it to occur then our claim amounts to the assertion that the maximum possible information is gained in this instance. Calculating that information we find,

$$
\begin{aligned}
\{\hat{\mathbf{r}}, \Phi_N\} &\leq \int \frac{(N+1)}{4\pi}\left(\frac{1+\hat{\mathbf{a}}\cdot\hat{\mathbf{r}}}{2}\right)^N \log\left[(N+1)\times\left(\frac{1+\hat{\mathbf{a}}\cdot\hat{\mathbf{r}}}{2}\right)^N\right]d\hat{\Omega}_{\hat{\mathbf{r}}} \\
&= \log(N+1) + N(N+1)\int\left(\frac{1+\hat{\mathbf{a}}\cdot\hat{\mathbf{r}}}{2}\right)^N \log\left(\frac{1+\hat{\mathbf{a}}\cdot\hat{\mathbf{r}}}{2}\right)d\hat{\Omega}_{\hat{\mathbf{r}}} \\
&= \log(N+1) + N(N+1)\int_0^1 u^N \log u\, du \\
&= \log(N+1) + N(N+1)\times\frac{-1}{(N+1)^2} \\
&= \log N - 1 + \log(1+1/N).
\end{aligned}
\tag{5.103}
$$

The last term goes to zero for large N, so it follows that the asymptotic bound on the Correlation Information is

$$
\{\hat{\mathbf{r}}, \Phi_N\} \leq \log N - 1.
\tag{5.104}
$$

Comparing this with eq.(5.99) it is clear that the uniform measurement set is the optimal apparatus. Interestingly, for large N, such a measurement set behaves as though one had fortuitously selected the correct direction for a singlet measurement.

This result is important in that it closes the gap as to any possible gains that might be achieved through Adaptive Inference. Such a procedure would involve using the accumulated data to appropriately modify the choice of bases during the course of the experiment. We have shown that there is no gain to be had from this elaboration. Indeed a single fixed automaton program, that is optimised over all possible inputs, achieves maximum extraction of information.

Given the relative closeness of the icosahedral set to the upper bound, we may suggest that use of this represents the optimal practical choice. It requires only 10 different analyser settings. All of these are to be used equally and inversion of the data is particularly simple.

103

In conclusion we can state that the quantum probability rule limits the asymptotic gain of information in N trials to be not greater than

$$\log N - 1.$$

This has physical consequences because spin states are one of the few ways to communicate orientation across space[50]. Pump–priming a stellar communication channel based upon observing photon polarisation would require the receipt of N identically polarised photons so as to be able acquire an error rate less than $1/N$ per subsequent signal symbol (photon).[11] An infinite number of priming photons would need to be observed in order to reduce the error rate to zero. However, after N have been transmitted the information carrying capacity of the channel becomes

$$C = 1 - 1/N \log_2(1/N) + (1 - 1/N) \log_2(1 - 1/N) \text{ bits.}$$

For large N this approaches one bit per photon as expected. Once the channel is primed this information carrying capacity can be used to signal, but cannot further reduce the error rate. That is, it cannot tell further information about the transmitter's orientation.

5.10   The inferred density matrix

Recall the formula for the inferred density matrix, eq.(5.7), page 48,

$$\begin{aligned}
\rho(\Phi_N) &= \int \frac{1}{2}(1 + \hat{\mathbf{r}} \cdot \vec{\sigma}) p(\hat{\mathbf{r}}|\Phi_N) \, d\hat{\Omega}_{\hat{\mathbf{r}}} \\
&= \frac{1}{2}(1 + \langle \hat{\mathbf{r}} \rangle \cdot \vec{\sigma}).
\end{aligned} \tag{5.105}$$

In §5.5 we calculated a series expression for the prior probabilities, $p(\Phi_N)$, of each outcome. A little thought shows that if we augment the set of vectors

$$\Phi_N \equiv \{\hat{\mathbf{a}}_i\}_{i=1}^N,$$

where the outcome signs have been absorbed, with the vector of Pauli matrices, $\vec{\sigma}$, so as to define

$$\Phi_N^*[\sigma_k] \equiv \{\hat{\mathbf{a}}_i\}_{i=1}^N \cup \{\sigma_k\}, \; k = 1, 2, 3;$$

---

[11] For an explanation of the concepts of error rate and channel capacity see [51, p.276].

then, in terms of the existing formula for $p(\Phi_N)$, we may write

$$\rho(\Phi_N) = \frac{1}{2}\left(1 + \frac{\sum_{k=1}^{3}\{p(\Phi_N^*[\sigma_k]) - p(\Phi_N^*[0])\}}{p(\Phi_N)}\right). \tag{5.106}$$

One use of the formula is required to calculate the denominator, then we consider each Pauli matrix as though it were a measurement outcome (so we treat $\sigma_{\hat{x}}$ like $\hat{x}$), and calculate the three components of the density matrix in terms of these. The subtraction of terms involving augmentation with the zero vector, 0, ensures that we include only those parts of the series expression for $p(\Phi_N)$ that contain the Pauli matrices.

For example, the first two inferred density matrices are given by,

$$\rho_1(\Phi_1) = \frac{1}{2}\left(1 + \frac{1}{3}\hat{a}_1 \cdot \vec{\sigma}\right) \tag{5.107}$$

$$\rho_2(\Phi_2) = \frac{1}{2}\left(1 + \frac{1/3(\hat{a}_1 + \hat{a}_2)}{(1 + 1/3\hat{a}_1 \cdot \hat{a}_2)} \cdot \vec{\sigma}\right). \tag{5.108}$$

To calculate the asymptotic density matrix, recall from eq.(5.75) that the inferred distribution for the unknown state $\hat{r}$ is given in terms of the observed data $\vec{X}$ by:

$$\begin{aligned} G(\vec{X}, \hat{r}) &\simeq 2N\{\det A_{\perp}\}^{+1/2} \\ &\times \exp -\frac{N}{2}\{(\hat{r} - \hat{s}_0)A_{\perp}(\hat{r} - \hat{s}_0)\}. \end{aligned} \tag{5.109}$$

From this we need to calculate

$$\langle \hat{r} \rangle \equiv \int \hat{r}\, G(\vec{X}, \hat{r})\, d\hat{\Omega}_{\hat{r}}. \tag{5.110}$$

It is only necessary to evaluate this for the asymptotically preferred outputs. These are outputs such that

$$\vec{X} = (\hat{a}_1 \cdot \hat{s}_0, \ldots, \hat{a}_m \cdot \hat{s}_0),$$

with $\hat{s}_0$ the maximum of the inferred distribution.

Local to this maximum we may set

$$\hat{r} \simeq \left(x, y, 1 - 1/2(x^2 + y^2)\right),$$

105

where the z-axis is aligned along $\hat{s}_0$. Then,

$$G(\vec{X}_0, \hat{r}) \simeq 2N \{\det \mathbf{A}_\perp\}^{+1/2}$$
$$\times \exp -\frac{N}{2}\{(x,y)\mathbf{A}_\perp(x,y)^T\} \qquad (5.111)$$

and

$$\langle \hat{r} \rangle \simeq 2N \{\det \mathbf{A}_\perp\}^{+1/2}$$
$$\times \int_{-\infty}^{\infty}\int_{-\infty}^{\infty} \left(x, y, 1 - 1/2(x^2 + y^2)\right) e^{-N/2\{(x,y)\mathbf{A}_\perp(x,y)^T\}} \, dx\,dy/4\pi.$$
$$= \left(0, 0, 1 - \{4N \det \mathbf{A}_\perp(\hat{s}_0)\}^{-1}\right). \qquad (5.112)$$

Clearly, as $N \to \infty$, the asymptotic density matrix approaches the pure state,

$$\rho = \frac{1}{2}(1 + \hat{s}_0 \cdot \vec{\sigma}).$$

It does so for any choice of apparatus which spans $\mathbf{R}^3$. Recall that,

$$0 < 4\det \mathbf{A}_\perp(\hat{s}_0) \leq 1.$$

Therefore, the above result may be alternatively interpreted as stating that the polarisation of the inferred density matrix must satisfy:

$$0 <\| \hat{s}_0 \|\leq 1 - 1/N.$$

Earlier, we intimated that there was a connection between the quantum mechanical entropy, S, and the Correlation Information. Employing convexity of the entropy function, it is clear that:

$$\log 2 \geq S \geq -(1 - 1/N)\log(1 - 1/N) - 1/N \log(1/N),$$

after N–trials. This is the entropy of the inferred density matrix. If we divide the upper bound to the Correlation Information after N–trials, by the number of systems observed one obtains

$$1/N\{\hat{r}, \Phi_N\} \leq 1/N \log N - 1/N.$$

This is the average Correlation Information acquired per system analysed. Hence, approximately:

$$\frac{\{\hat{r}, \Phi_N\}}{N} \leq \frac{\log N}{N} \leq S.$$

We should not attempt to invest too much significance in this. The entropy, S, has no connection with that created in the experiment. This equation simply expresses the fact that there is a correlation between the localisation of a normalised distribution upon the sphere, measured by $\{\hat{\mathbf{r}}, \Phi_N\}$, and the length of its vector–valued mean, measured by S, with these two related to the same determinant, $\det \mathbf{A}_\perp(\hat{\mathbf{s}}_0)$.

Having thoroughly explored application of the theory to systems with two states, we shall now explore the generalisation to systems of aribtrary finite dimensionality. Central to this generalisation is the development of an integration measure to play the role of the spherical integrations used in this chapter. There is a natural choice, which corresponds to a uniform prior. With this choice we are able to develop the theory for uniform measurement sets and to obtain an upper bound to the Correlation Information.

Chapter 6

General OSDP

6.1 Ray integration measure

It should be clear that the only difficulty that hinders generalisation is the loss of the sphere representation for the space of rays. From §1.3, on the geometry of density matrices, we know that there is no problem with the existence of probability measures. A vector representation of projectors is still possible but because of the "holes" in S_{d^2-2} it is unhelpful to follow this route. If one could perform the integration over rays as one over this hypersphere then the generalisation of previous results would amount to some trivial numerical factors.

Rather, we must retain the complex vector representation of rays and carry out the summation over all possible states as an integration over all possible complex vectors. This can be done in several ways. We illustrate with two examples that are tailored to particular types of integrand.

Of interest are integrands that are linear in $|\psi>$ and antilinear in $<\psi|$. That is to say ψ will appear in functions as $<\psi|\phi><\phi|\psi>$ for some projector $|\phi><\phi|$. Since the space is finite–dimensional we should observe that kets can equally well be thought of as normalised d–dimensional complex column vectors $\mathbf{z}$, whence the corresponding bra is the row vector $\mathbf{z}^\dagger$. So both states and elements of a measurement basis translate to the dyadics of such vectors, written $\mathbf{z}\mathbf{z}^\dagger$ (a projector).[1]

---

[1]This notation is typographically more convenient. The substitutions: $|\psi> \leftrightarrow \mathbf{z}$ and $<\psi| \leftrightarrow \mathbf{z}^\dagger$ may be safely performed at any stage.

Two possible classes of integrand arise. Those that satisfy the scaling law:

$$\exists \, k \in \mathbf{Z} \vdash F(\lambda \mathbf{z}, \lambda^* \mathbf{z}^\dagger) = (\lambda \lambda^*)^k F(\mathbf{z}, \mathbf{z}^\dagger) \; \forall \lambda \in \mathbf{C}, \qquad (6.1)$$

and those that do not.

To apply the theory of inference that has been developed we require a uniform integration measure over objects $\mathbf{z}\mathbf{z}^\dagger$. Recall that these stand in one–one correspondence with the rays of the Hilbert space, denoted $\bar{\mathbf{z}}$.

For integrands in the first class a uniform ray measure is obtained by taking the following expression:

$$\int F(\mathbf{z}, \mathbf{z}^\dagger) \, d\hat{\Omega}_{\bar{\mathbf{z}}} = \frac{1}{\mathcal{N}} \int F(\mathbf{z}, \mathbf{z}^\dagger) e^{-\mathbf{z}^\dagger \mathbf{z}} \, d\mathbf{z}, \qquad (6.2)$$

where $d\hat{\Omega}_{\bar{\mathbf{z}}}$ is used for consistency of notation. This denotes a normalised uniform sum over all rays $\bar{\mathbf{z}}$. Note that the invariance of the above expression under arbitrary unitary change of variables expresses the uniformity of the measure.

The integration on the right–hand side may be carried out in terms of one over the complex components of vector $\mathbf{z}$ as:

$$\int (\bullet) \, d\mathbf{z} \equiv \int_{-\infty}^{\infty} \cdot^{2d} \cdot \int_{-\infty}^{\infty} (\bullet) \prod_{l=1}^{d} dx_l dy_l. \qquad (6.3)$$

Then the normalisation is given in terms of the scaling parameter k by

$$\mathcal{N} \equiv \int (\mathbf{z}^\dagger \mathbf{z})^k e^{-\mathbf{z}^\dagger \mathbf{z}} \, d\mathbf{z}; \qquad (6.4)$$

which is

$$
\begin{aligned}
\mathcal{N} &= \int_{-\infty}^{\infty} \cdot^{2d} \cdot \int_{-\infty}^{\infty} \left(\sum_{l=1}^{d} x_l^2 + y_l^2 \right)^k \prod_{l=1}^{d} e^{-\left(x_l^2 + y_l^2 \right)} \, dx_l dy_l \\
&= \left[(-1)^k \frac{d^k}{d\lambda^k} \int_{-\infty}^{\infty} \cdot^{2d} \cdot \int_{-\infty}^{\infty} \prod_{l=1}^{d} e^{-\lambda \left(x_l^2 + y_l^2 \right)} \, dx_l dy_l \right]_{\lambda=1} \\
&= \left[(-1)^k \frac{d^k}{d\lambda^k} \left(\frac{\pi}{\lambda} \right)^d \right]_{\lambda=1} \\
&= \pi^d \times \frac{(d + k - 1)!}{(d - 1)!}. \qquad (6.5)
\end{aligned}
$$

Therefore, we find that

$$\int F(\mathbf{z}, \mathbf{z}^\dagger)\, d\hat{\Omega}_{\bar{z}} = \frac{(d-1)!}{\pi^d (d+k-1)!} \int F(\mathbf{z}, \mathbf{z}^\dagger) e^{-\mathbf{z}^\dagger \mathbf{z}}\, d\mathbf{z}. \qquad (6.6)$$

In general the scaling relation is not satisified and one must use the following delta function form:

$$\int (\bullet)\, d\hat{\Omega}_{\bar{z}} = \int (\bullet)\, \delta(1 - \mathbf{z}^\dagger \mathbf{z})\, d\mathbf{z}$$

$$= \frac{1}{\mathcal{N}} \int_{-\infty}^{\infty} e^{-ik}\, dk \int_{-\infty}^{\infty} \overset{2d.}{\cdots} \int_{-\infty}^{\infty} (\bullet) \prod_{l=1}^{d} e^{+ik(x_l^2 + y_l^2)}\, dx_l dy_l. \qquad (6.7)$$

The normalisation is

$$\mathcal{N} = \int_{-\infty}^{\infty} e^{-ik}\, dk \int_{-\infty}^{\infty} \overset{2d.}{\cdots} \int_{-\infty}^{\infty} \prod_{l=1}^{d} e^{+ik(x_l^2 + y_l^2)}\, dx_l dy_l$$

$$= \int_{-\infty}^{\infty} \left(\frac{\pi}{-ik}\right)^d e^{-ik}\, dk$$

$$= 2\pi \frac{\pi^d}{(d-1)!}. \qquad (6.8)$$

Here use has been made of the result

$$\frac{1}{2\pi} \int_{-\infty}^{\infty} \left(\frac{1}{-ik}\right)^w e^{-ik}\, dk = \frac{1}{\Gamma(w)}, \qquad (6.9)$$

with the contour closed in the lower half plane. Further to this we note the two useful integrals:

$$\int_{-\infty}^{\infty} e^{+ikx^2}\, dx = \left(\frac{+i\pi}{k}\right)^{1/2} \qquad (6.10)$$

$$\int_{-\infty}^{\infty} e^{-ikx^2}\, dx = \left(\frac{-i\pi}{k}\right)^{1/2}. \qquad (6.11)$$

Substituting eq.(6.8) into the definition eq.(6.7) yields

$$\int (\bullet)\, d\hat{\Omega}_{\bar{z}} = \frac{(d-1)!}{\pi^d} \times \frac{1}{2\pi} \int_{-\infty}^{\infty} e^{-ik}\, dk \int_{-\infty}^{\infty} \overset{2d.}{\cdots} \int_{-\infty}^{\infty} (\bullet) \prod_{l=1}^{d} e^{+ik(x_l^2 + y_l^2)}\, dx_l dy_l.$$
$$(6.12)$$

110

So we have two realisations of the uniform ray measure. Note that the second choice is more general and so it is a matter of convenience as to which is used when the scaling relation eq.(6.1) is satisfied.

Having established the means by which the integration over possible states is to be carried out we can now follow the pattern of development pursued for two–state systems. Of course any formulæ developed here will reduce to those of the last chapter upon substitution of $d = 2$. However, we shall not be concerned with verifying this explicitly.

The major question addressed is the generalisation of the result given in §5.8 for the two–state uniform measurement set. We shall not study the general asymptotics, as bounding the Correlation Information is of greater theoretical interest. This we do with the appropriate generalisation of the argument developed in §5.9. In addition the general series expression for prior probabilities is obtained which represents a solution in principle to the state–inference problem for arbitrary measurement sets. Before embarking upon this program we discuss the nature of generalised measurement bases.

6.2   General measurement bases

Recalling the general formulation of the *OSDP* given in §4.2, p. 42, we can now rapidly develop the necessary general results. The only point that is essentially new concerns the specification of measurement bases. Now there are more than two outcomes. Therefore, in the kth of N, trials one of d projectors:

$$P_j^k \equiv |\phi_j^k><\phi_j^k| \quad j \in [1,d]; \; k \in [1,N],$$

from the kth measurement basis is "seen" as an outcome.

The signs, σ_k, of the last chapter, provided a simple way to label the outcomes. This feature is lost now. Rather, it is necessary to retain the individual projector status of each alternative outcome. Recall from §1.3 that the d projectors appear geometrically as a d–vertex equi–pyramid in the space $\mathbf{R}^{d^2-1}$. The closure relation,

$$\sum_{j=1}^{d} P_j^k = \mathbf{I}, \tag{6.13}$$

that must exist for these to be a complete set, expresses itself in the fact that the barycentre of such an equi–pyramid is the matrix $1/d\,\mathbf{I}$. Of course this lies at the centre of the hypersphere S_{d^2-2} in the plane of pre–density matrices.

In the general case not all orientations of equi–pyramids are bases. This is because there are non–positive regions to the surface of the Generalised Poincaré sphere. These cannot correspond to the, necessarily idempotent, projectors of a basis.

As a particular example of such restrictions; we already know that the maximum number of mutually unbiased bases[2] is $d + 1$. Furthermore this is achieved for prime dimensional spaces but would appear impossible for composite d. Ivanovič, [30], has given the unitary operators corresponding to such mutually unbiased bases for prime d. We shall use these to explore the possibilty that such bases achieve for general prime d the equi–probable results for $N = d+1$, as has already been observed in the two–state case. This corresponds to realising maximum possible information for up to $N = d + 1$ trials.

6.3 Simplification of the Correlation Information formula

If we consider the general equation eq.(4.20) given on page 46 and examine the argument used to simplify the first term for two–state systems, see eq.(5.8) through eq.(5.11), then one finds that the only essential property used was the closure relation eq.(6.13). This is retained in the general case, as are the necessary symmetry properties of the integration measure when passage is made to $d\hat{\Omega}_{\psi}$. Therefore, we may assert that the general result is:

$$\{\psi, \Phi_N\} = C_N + H(\Phi_N), \tag{6.14}$$

where

$$C_N = d \times N \times C_0 \tag{6.15}$$

$$C_0 = \int <\psi|\phi><\phi|\psi> \log[<\psi|\phi><\phi|\psi>]\, d\hat{\Omega}_{\psi} \tag{6.16}$$

are the constants corresponding to eq.(5.11). Of course $H(\Phi_N)$ is the usual Information Entropy defined upon the prior probabilities of outcomes:

$$\Phi_N \equiv \{\phi_{j_k}^k\}_{k=1}^N = \{P_{j_k}^k\}_{k=1}^N.$$

---

[2]In the sense of Wootters [29].

The expanded indices $j_k \in [1, d]$ will be used occasionally.

The probabilities can be made explicit as

$$p(\Phi_N) = \int \prod_k^N <\psi|P_{j_k}^k|\psi> \, d\hat{\Omega}_{\vec{\psi}}. \tag{6.17}$$

Our task is now to calculate both eq.(6.15) and eq.(6.17). We tackle the constant C_0 first.

Notice that symmetry considerations ensure that the result is independent of ϕ. Furthermore the integrand does not scale so we must use the delta function measure. Choosing a convenient ϕ and substituting the integrand into eq.(6.12) gives

$$\int <\psi|\phi><\phi|\psi> \log[<\psi|\phi><\phi|\psi>] \, d\hat{\Omega}_{\vec{\psi}} = \frac{(d-1)!}{\pi^d} \times \frac{1}{2\pi}$$

$$\times \int_{-\infty}^{\infty} \left(\frac{+i\pi}{k}\right)^{d-1} e^{-ik} \, dk \int_{-\infty}^{\infty} \int_{-\infty}^{\infty} (x_1^2 + y_1^2) \log(x_1^2 + y_1^2) e^{+ik(x_1^2 + y_1^2)} \, dx_1 dy_1.$$

Here ϕ has been chosen as the first basis element and the $d-1$ pairs of elementary integrals of form eq.(6.10) have been carried through. We can now drop subscripts on x, y and change variables to polar coordinates $r^2 = x^2 + y^2$ to obtain

$$C_0 = \frac{(d-1)!}{\pi^d} \times \frac{1}{2\pi} \int_{-\infty}^{\infty} \left(\frac{+i\pi}{k}\right)^{d-1} e^{-ik} \, dk \int_0^{\infty} r^2 \log r^2 \, e^{+ikr^2} \, 2\pi r \, dr. \tag{6.18}$$

Another change of variables to $u = r^2$ and cancellation of factors π yields

$$C_0 = (d-1)! \times \frac{1}{2\pi} \int_{-\infty}^{\infty} \left(\frac{+i}{k}\right)^{d-1} e^{-ik} \, dk \int_0^{\infty} u \log u \, e^{+iku} \, du. \tag{6.19}$$

Now to solve this observe that the integral for C_0 is certainly well defined for it is one over a compact manifold with $<\psi|\phi><\phi|\psi> \in [0,1]$. Therefore, we can afford to be optimistic about things. Defining

$$F_m(u) \equiv \frac{1}{2\pi} \int_{-\infty}^{\infty} \left(\frac{+i}{k}\right)^m e^{-ik(1-u)} \, dk, \tag{6.20}$$

we can identify F as a generalised theta function. The function F_1 is a step function which is zero for $u > 1$ whilst F_0 is an ordinary delta function. The

113

properties of this distribution may be summarised as follows:

$$\begin{aligned}
\frac{d}{du}F_m(u) &= (-1)F_{m-1}(u)\\
F_1(u) &= \Theta_L(1-u)\\
F_0(u) &= \delta(1-u).
\end{aligned}\qquad (6.21)$$

With this notation the original integral becomes

$$C_0 = (d-1)! \times \int_0^\infty F_{d-1}(u)\,u\log u\,du. \qquad (6.22)$$

Integration by parts is suggestive. To do this we require the order $d-1$ antiderivative of $u\log u$. To this end define

$$[f(u)]^1 \equiv \int^u f(s)ds$$

and

$$[f]^m = [[f]^{m-1}].$$

Then we claim (for $m \geq 2$):

$$[u\log u]^{m-1} = \frac{u^m}{m!}\log u - \frac{1}{m!}(\frac{1}{2}+\ldots+\frac{1}{m})u^m. \qquad (6.23)$$

This we prove by induction. Notice that

$$[u\log u]^1 = \frac{u^2}{2}\log u - \frac{1}{4}u^2$$

so the claim is true for $m=2$. To complete the proof we calculate

$$\begin{aligned}
[[f]^{m-1}] &= \int^u \left\{\frac{s^m}{m!}\log s - \frac{1}{m!}(\frac{1}{2}+\ldots+\frac{1}{m})s^m\right\}ds\\
&= \frac{u^{m+1}}{(m+1)!}\log u - \frac{1}{m+1}\frac{u^{m+1}}{(m+1)!}\\
&\quad -\frac{1}{(m+1)!}(\frac{1}{2}+\ldots+\frac{1}{m})u^{m+1}\\
&= \frac{1}{(m+1)!}u^{m+1}\log u - \frac{1}{(m+1)!}(\frac{1}{2}+\ldots+\frac{1}{m+1})u^{m+1}\\
&= [f]^m,
\end{aligned}$$

114

as required. Performing one integration by parts in eq.(6.22), and making use of eq.(6.21) we find

$$C_0/(d-1)! = F_{d-1}(u)[u\log u]|_0^\infty - \int_0^\infty (-1)F_{d-2}(u)\,[u\log u]\,du. \qquad (6.24)$$

The first term vanishes at zero because of the log term and at infinity because the distribution F vanishes above $u = 1$. Therefore, taking $d-1$ iterations of this procedure yields

$$C_0 = (-1)^{d-1}(d-1)! \times \int_0^\infty (-1)^{d-1}\delta(1-u)\,[u\log u]^{d-1}\,du. \qquad (6.25)$$

Substituting for $[u\log u]^{d-1}$ from eq.(6.23) and evaluating the expression at $u = 1$ we have

$$C_0 = -(d-1)! \times \frac{1}{d!}(\frac{1}{2} + \ldots + \frac{1}{d}). \qquad (6.26)$$

Notice that at $d = 2$ we recover the two–state result $C_0 = -1/4$. Calculating C_N from eq.(6.15) gives the final expression

$$C_N = -N \times (\frac{1}{2} + \ldots + \frac{1}{d}). \qquad (6.27)$$

Thus we arrive at the simplified expression

$$\{\psi, \Phi_N\} = -N \times (\frac{1}{2} + \ldots + \frac{1}{d}) + H(\Phi_N). \qquad (6.28)$$

Now we turn to the calculating the probabilities given by eq.(6.17).

6.4   General probability formula

In z notation eq.(6.17) is

$$p(\Phi_N) = \int \prod_k^N (z^\dagger P_{j_k}^k z)\,d\hat\Omega_{\hat z}. \qquad (6.29)$$

It is helpful to recast this in tensor notation. Therefore, define the measurement tensor:

$$M_{l_1\ldots l_N}^{l_1'\ldots l_N'}(\Phi_N) \equiv \prod_{k=1}^N P_{j_k}^k(l_k, l_k'), \qquad (6.30)$$

115

where $P_{j_k}^k(l_k, l_k')$ are the matrix elements of projector j_k from the kth basis. Similarly define a state tensor:

$$\Psi_{l_1'\cdots l_N'}^{l_1\cdots l_N}(\bar{z}) \equiv \prod_{k=1}^{N} z_{l_k}^\dagger z_{l_k'}. \tag{6.31}$$

Then eq.(6.17) becomes

$$p(\Phi_N) = M_{l_1\cdots l_N}^{l_1'\cdots l_N'}(\Phi_N) \times \int \Psi_{l_1'\cdots l_N'}^{l_1\cdots l_N}(\bar{z}) \, d\hat{\Omega}_{\bar{z}} \tag{6.32}$$

with contraction over repeated indices implicit.

Having pulled out the Φ_N dependence it is clear that we have only to calculate what must be a constant tensor with certain symmetry properties. For instance we can certainly permute both upper and lower indices because the integration measure is uniform. Furthermore, considered as a function of z it satisfies the scaling property of eq.(6.1).

The following possibility suggests itself

$$\int \Psi_{l_1'\cdots l_N'}^{l_1\cdots l_N}(\bar{z}) \, d\hat{\Omega}_{\bar{z}} \propto S_{l_1'\cdots l_N'}^{l_1\cdots l_N} \tag{6.33}$$

where

$$S_{l_1'\cdots l_N'}^{l_1\cdots l_N} \equiv \begin{vmatrix} \delta_{l_1'}^{l_1} & \delta_{l_1'}^{l_2} & \cdots & \delta_{l_1'}^{l_N} \\ \delta_{l_2'}^{l_1} & \delta_{l_2'}^{l_2} & \cdots & \delta_{l_2'}^{l_N} \\ \vdots & \vdots & & \vdots \\ \delta_{l_N'}^{l_1} & \delta_{l_N'}^{l_2} & \cdots & \delta_{l_N'}^{l_N} \end{vmatrix}_+ \tag{6.34}$$

is $N!$ times the symmetrisation operator[52] and $|\ \ |_+$ denotes the permanent (determinant with all signs positive). To verify this observe that in virtue of the scaling property we find upon substitution into eq.(6.6) that

$$\int \Psi_{l_1'\cdots l_N'}^{l_1\cdots l_N}(\bar{z}) \, d\hat{\Omega}_{\bar{z}} = \frac{(d-1)!}{\pi^d (d+N-1)!} \int \prod_{k=1}^{N} (z_{l_k}^\dagger z_{l_k'}) \times e^{-z^\dagger z} \, d\hat{\Omega}_{\bar{z}}. \tag{6.35}$$

Now we are interested in the value of this for particular choices of the indices (l_k, l_k'). A little experimentation shows that the upper indices must be a permutation of the lower ones or the answer will be zero. The integral is

116

independent of the actual value of the indices. It only depends on the number which are the same. Therefore, count up how many times each of the d components of z_i appears and call this r_i, with

$$\sum_{i=1}^{d} r_i = N.$$

Then for those indices (l_k, l'_k) in the above class we have:

$$\int \Psi_{l'_1 \cdots l'_N}^{l_1 \cdots l_N}(\bar{z}) \, d\hat{\Omega}_{\bar{z}} = \frac{(d-1)!}{\pi^d (d+N-1)!} \int \prod_{i=1}^{d} (z_i^* z_i)^{r_i} e^{-\mathbf{z}'\mathbf{z}} \, d\hat{\Omega}_{\bar{z}}. \tag{6.36}$$

The integral now splits up with a generic factor being

$$\int_{-\infty}^{\infty} \int_{-\infty}^{\infty} (x^2 + y^2)^{r_i} \times e^{-(x^2 + y^2)} \, dx \, dy = \pi \cdot r_i!.$$

Thus we are left with

$$\int \Psi_{l'_1 \cdots l'_N}^{l_1 \cdots l_N} \, d\hat{\Omega}_{\bar{z}} = \frac{(d-1)!}{(d+N-1)!} \prod_{i=1}^{d} r_i! \tag{6.37}$$

provided that the permutation condition is satisfied. One can then convince oneself that the permanent generates the necessary factor $r_1! \cdots r_d!$ because for fixed lower indices the upper ones run through all permutations of these and zeros will result from the Kronecker deltas unless they match. For a given partition of N into d powers r_i there are precisely

$$\prod_{i=1}^{d} r_i!$$

non-zero terms in the permanent expansion. Hence we have the final result

$$\int \Psi_{l'_1 \cdots l'_N}^{l_1 \cdots l_N}(\bar{z}) \, d\hat{\Omega}_{\bar{z}} = \frac{(d-1)!}{(d+N-1)!} S_{l'_1 \cdots l'_N}^{l_1 \cdots l_N}. \tag{6.38}$$

The probability must then be given by the contraction of this with the measurement tensor, which is

$$p(\Phi_N) = \frac{(d-1)!}{(d+N-1)!} \, M_{l_1 \cdots l_N}^{l'_1 \cdots l'_N}(\Phi_N) \times S_{l'_1 \cdots l'_N}^{l_1 \cdots l_N}. \tag{6.39}$$

117

An alternative route to this result is to use the generating function

$$p(\Phi_N) = \left[\frac{\partial}{\partial\lambda_1}\cdots\frac{\partial}{\partial\lambda_N}G(\lambda_1,\ldots,\lambda_N)\right]_{\vec{\lambda}=0}. \qquad (6.40)$$

Where G is defined in terms of the matrix

$$\mathbf{M}(\vec{\lambda}) \equiv \sum_{k=1}^{N}\lambda_k\mathbf{P}_{j_k}^{k}$$

as

$$\begin{aligned}G(\lambda_1,\ldots,\lambda_N) &\equiv \frac{(d-1)!}{(d+N-1)!}\int\exp\left\{-\mathbf{z}^\dagger[\mathbf{I}-\mathbf{M}(\vec{\lambda})]\mathbf{z}\right\}\,d\mathbf{z}\\ &= \frac{(d-1)!}{(d+N-1)!}\det[\mathbf{I}-\lambda_k\mathbf{P}_{j_k}^{k}]^{-1}.\end{aligned} \qquad (6.41)$$

To recover the series one uses the standard trick[47, p.187],

$$\frac{1}{\det[\mathbf{I}-\mathbf{M}]} = \exp\left\{-\operatorname{Tr}\log[\mathbf{I}-\mathbf{M}]\right\},$$

from which a power series can be developed. Proceeding in much the same way as we did when using the generating function for two–states enables reproduction of the formula eq.(6.39).

In the same manner as before one can use the prior probability formula derived above to obtain the exact inferred density matrix.

Now we turn to examination of some numerical experiments conducted upon mutually unbiased bases in prime spaces. These were only possible for small values of N and d, because of the very rapid growth in computation time. Attempting to repeat the simulated annealing experiments is out of the question.

6.5 Mutually unbiased bases: numerical experiments

Mutually unbiased bases for prime $d = q \geq 2$ are listed in [29, p.512] as:

$$B_l^{0m} = \delta_{ml}, \qquad (6.42)$$

118

$$B_l^{km} = \frac{1}{\sqrt{q}} \exp\left[\frac{2\pi i}{q} k(m+l)^2\right], \qquad (6.43)$$

$$B_l^{qm} = \frac{1}{\sqrt{q}} \exp\left[\frac{2\pi i}{q} ml\right], \qquad (6.44)$$

where the index $m \in [1, q]$ labels rays from a given basis, $l \in [1, q]$ labels the components of such a ray in the reference basis (first one) and $k \in [1, q-1]$ labels the $q-1$ bases generated by the middle formula for different values of k. In all there are $q+1$ bases. Using elementary properties of Gauss sums, one can verify by explicit calculation that they satisfy the unbiased property.

In the case of two–states we found that it was possible to maintain equally probable outcomes for up to three measurements. It turns out that this property is lost for higher dimensions.

We tested the formula for prior probabilities for the cases $d = 3$ and $d = 5$. It is possible to show analytically that for two observations equally probable outcomes are always maintained. However, for $N \geq 3$ equi-probability is lost.

A computer program was written to generate all possible measurement tensors and form the appropriate contraction. For N trials with dimension d there are d^N of these so it was not possible to carry the calculation very far. The Correlation Information was calculated and a list formed of the various outcome probabilities. The results are collected in table(6.1). It is interesting that the property of equality of outcome probabilities fails to generalise in higher dimensions for N greater than two. The reason for this lies in the presence of terms within the expansion of the contracted measurement tensor that involve traces of the product of odd numbers of projectors. The mutually unbiased property is not sufficient to guarantee that these will be constant for all possible selections of rays from the different unbiased bases.

6.6 Uniform measurement set

Asymptotic analysis is somewhat more complicated in this case. However, one might conjecture a growth in Correlation Information that goes like

$$(d-1)\log N$$

in terms of the dimensionality and number of trials. This is because the pure states form a dimension $2d - 2$ submanifold in $\mathbf{R}^{d^2-1}$. One expects that the

119

dim	trials	information	probability	number
3	2	0.530558	0.111111	9
	3	0.793292	0.038889	9
			0.033333	18
	4	1.049633	0.007407	9
			0.012963	72
5	2	0.652209	0.040000	25
	3	0.975220	0.006767	25
			0.008308	100
	4	1.297434	0.001333	25
			0.001358	200
			0.001690	200
			0.001784	200

Table 6.1: Tabulated values of probabilities, with the number of times they occur, and the Correlation Information for $d = 3$ and $d = 5$ mutually unbiased bases. The calculation is for numbers of trials, N, less than or equal to $d + 1$, the maximum number of possible unbiased bases. Only prime d were tested and the calculation proved prohibitive beyond $d = 5$ and $N = 4$. The calculation was double precision.

asymptotic location of the state will be Gaussian around some point of this manifold. Integrating in the tangent space to this point should pick up a factor of $\sqrt{N}$ for each linear dimension (for example: when $d = 2$ we get $1/2 \times (2d - 2) = d - 1$).

Rather than pursue this further it is more productive to concentrate upon bounding the Correlation Information.

Taking the natural generalisation of the argument given in §5.8 we expect the following result for generalised uniform measurement sets:

$$p(\phi|\psi) = \frac{1}{\mathcal{N}} \exp Nd \left\{ \int <\phi|\omega><\omega|\phi> \log \left[<\psi|\omega><\omega|\psi> \right] \right\} d\hat{\Omega}_{\hat{\omega}}. \quad (6.45)$$

This applies for large N as an effective correlation between the near–pure state, ϕ, that one infers and the true input state, ψ. Law–of–large–numbers behaviour will ensure that there is vanishing measure for other results as $N \to \infty$.

In order to evaluate this integral we seek to recast it in terms of that which we already know from the C_0 calculation, eq.(6.16).

Choose a basis with d^{th} element aligned along $|\psi>$. That is, substitute:

$$|\omega> = z = (x + iy)|\psi> + \sum_{j=1}^{d-1}(x_j + iy_j)|j> .$$

Thus the exponent of eq.(6.45) becomes

$$\begin{aligned} E &= Nd \times \int | <\phi|\psi> |^2(x^2 + y^2)\log(x^2 + y^2)\, d\hat{\Omega}_{\hat{z}} \\ &\quad + Nd \times \sum_{j=1}^{d-1} \int | <j|\psi> |^2(x_j^2 + y_j^2)\log(x^2 + y^2)\, d\hat{\Omega}_{\hat{z}}. \quad (6.46) \end{aligned}$$

Here an initial simplification has been made by observing that moments involving cross terms between different basis coordinates will vanish. Now the first integral may be identified from eq.(6.15) as

$$Nd \times | <\phi|\psi> |^2 \times C_0. \quad (6.47)$$

For the second integral we may pull out the factor

$$| <j|\psi> |^2$$

121

and realise that the remaining integral is independent of index j through symmetry. Therefore, let us evaluate just one of them. Take the one for index $j = 1$. Explicitly this integral can be written as

$$\int (x_1^2 + y_1^2) \log(x^2 + y^2)\, d\hat{\Omega}_{\hat{z}} = \frac{(d-1)!}{\pi^d} \times \frac{1}{2\pi}$$
$$\times \int_{-\infty}^{\infty} \left(\frac{+i\pi}{k}\right)^{d-2} e^{-ik}\, dk$$
$$\int_{-\infty}^{\infty} \cdot\, \overset{4}{\cdot}\, \cdot \int_{-\infty}^{\infty} (x_1^2 + y_1^2) \log(x^2 + y^2) e^{+ik(x_1^2 + y_1^2 + x^2 + y^2)}\, dx_1 dy_1 dx dy.$$

As usual the coordinates which appear only in the exponent have been integrated out. Noting that

$$\int (x^2 + y^2) e^{+ik(x^2+y^2)}\, dx dy = -i \frac{d}{dk}\left(\frac{\pi i}{k}\right)$$
$$= \pi \left(\frac{i}{k}\right)^2,$$

the previous equation becomes:

$$\int (x_1^2 + y_1^2) \log(x^2 + y^2)\, d\hat{\Omega}_{\hat{z}} = \frac{(d-1)!}{\pi} \times \frac{1}{2\pi}$$
$$\times \int_{-\infty}^{\infty} \left(\frac{+i}{k}\right)^{d} e^{-ik}\, dk \int_{-\infty}^{\infty} \cdot\, \overset{4}{\cdot}\, \cdot \int_{-\infty}^{\infty} \log(x^2 + y^2) e^{+ik(x^2+y^2)}\, dx dy,$$

with some factors of π cancelled. Following an argument similar to that used to calculate C_0 from the point eq.(6.18) we can reduce this to

$$\int (x_1^2 + y_1^2) \log(x^2 + y^2)\, d\hat{\Omega}_{\hat{z}} = (d-1)! \times \int_0^{\infty} F_d(u) \log u\, du. \qquad (6.48)$$

Noticing the relation

$$[\log u]^1 = u \log u - u,$$

we find

$$[u \log u]^d = [u \log u]^{d-1} - [u]^{d-1}.$$

So the above integral is

$$\int_0^{\infty} F_d(u) \log u\, du = \int_0^{\infty} \delta(1-u)[u \log u - u]^{d-1}\, du. \qquad (6.49)$$

122

Now

$$\int_0^\infty \delta(1-u)[u]^{d-1}\, du = \frac{1}{d!}$$

and comparing eq.(6.49) to eq.(6.22) shows that

$$(d-1)! \int_0^\infty F_d(u) \log u\, du = C_0 - \frac{1}{d}. \qquad (6.50)$$

Returning to the second integral of eq.(6.46) we have the result

$$Nd \times \sum_{j=1}^{d-1} |<j|\psi>|^2 \times \left(C_0 - \frac{1}{d}\right).$$

However,

$$\sum_{j=1}^{d-1} |<j|\psi>|^2 = 1 - |<\phi|\psi>|^2,$$

so this can be rewritten as

$$Nd \times (C_0 - \frac{1}{d})(1 - |<\phi|\psi>|^2). \qquad (6.51)$$

Substituting eq.(6.51) and eq.(6.47) for the two integrals in eq.(6.46) we obtain the final result

$$
\begin{aligned}
E &= Nd \times \left\{ C_0|<\phi|\psi>|^2 + (C_0 - \frac{1}{d})(1 - |<\phi|\psi>|^2) \right\} \\
&= Nd \times \left\{ C_0 - \frac{1}{d} + \frac{1}{d}|<\phi|\psi>|^2 \right\} \\
&= N \times \left\{ dC_0 - 1 + |<\phi|\psi>|^2 \right\}. \qquad (6.52)
\end{aligned}
$$

To check this, compare it to the equation given in §5.8. For $d=2$, $C_0 = -1/4$ and

$$|<\phi|\psi>|^2 = \frac{1}{2}(1+\hat{s}\cdot\hat{r}).$$

Making these substitutions, the bracketed term in eq.(6.52) is

$$2 \times -1/4 - 1 + \frac{1}{2}(1+\hat{s}\cdot\hat{r}) = -1 + \frac{1}{2}\hat{s}\cdot\hat{r},$$

as required.

123

Returning to eq.(6.45) we can now substitute the result eq.(6.52), although there is no need to retain constant terms since these may be absorbed into the normalisation $\mathcal{N}$. Doing this we obtain

$$p(\phi|\psi) = \frac{1}{\mathcal{N}} \times \exp N \left\{ <\phi|\psi><\psi|\phi> \right\}. \tag{6.53}$$

This is a very interesting result. Choosing a uniform measurement set realises an apparatus whose state-space, X_A, is isomorphic to the system state-space, X_S. Furthermore, the correct large N effective correlation for such an apparatus is just the exponential of an appropriately scaled version of the quantum transition probability between $|\psi>$ and $|\phi>$. However, notice that $|\psi>$ is a *physical* state, whilst $|\phi>$ represents our *knowledge* of this state after N ensemble members have been measured.

To calculate the normalisation we must evaluate

$$\mathcal{N} = \int \exp N \left\{ <\phi|\psi><\psi|\phi> \right\} \, d\hat{\Omega}_{\hat{\phi}}. \tag{6.54}$$

Employing the delta function measure this is

$$\mathcal{N} = \frac{(d-1)!}{\pi^d} \times \frac{1}{2\pi}$$
$$\times \int_{-\infty}^{\infty} \left(\frac{+i\pi}{k} \right)^{d-1} e^{-ik} \, dk \int_{-\infty}^{\infty} \int_{-\infty}^{\infty} e^{(N+ik)(x^2+y^2)} \, dx dy.$$

One element of the expansion basis has been chosen to lie along $|\psi>$ and non-explicit variables have been integrated out. The apparent divergence of this integral is no problem because of ray-space compactness.

To do this integral we first rewrite the normalisation in F notation as

$$\frac{\mathcal{N}}{(d-1)!} = \int_0^{\infty} F_{d-1}(u) e^{Nu} \, du. \tag{6.55}$$

Now observe the following two properties of distribution F:

$$F_m(0) = \frac{1}{(m-1)!}, \tag{6.56}$$

$$\frac{d}{du} F_m(u) = -F_{m-1}(u). \tag{6.57}$$

124

The first comes from making use of the result eq.(6.9) to evaluate F at zero, whilst the second was listed previously in eq.(6.21).

We wish to apply integration by parts to eq.(6.55), but now care is needed to pick up a contribution from the leading integration by parts term evaluated at the lower limit of zero.

We now make the following conjecture:

$$\int_0^\infty F_m(u)e^{Nu}\,du = \frac{1}{N^m} \times \sum_{k=m}^\infty \frac{N^k}{k!}. \tag{6.58}$$

This shall be proved by induction. First at $m = 1$, eq.(6.58) becomes

$$\int_0^\infty F_1(u)e^{Nu}\,du = \int_0^1 e^{Nu}\,du = \frac{1}{N}\left[e^N - 1\right].$$

This is because $F_1(u)$ is a step function which is zero to the right of $u = 1$. The right–hand side agrees with that of eq.(6.58) so the conjecture is true for $m = 1$.

Now let us calculate the integral:

$$\begin{aligned}
\int_0^\infty F_{m+1}(u)e^{Nu}\,du &= F_{m+1}(u)\frac{e^{Nu}}{N}\Big|_0^\infty + \int_0^\infty F_m(u)\frac{e^{Nu}}{N}\,du \\
&= \left[0 - \frac{F_{m+1}(0)}{N}\right] + \frac{1}{N} \times \left[\frac{1}{N^m}\sum_{k=m}^\infty \frac{N^k}{k!}\right] \\
&= -\frac{1}{m!}\cdot\frac{1}{N} + \frac{1}{N} \times \left[\frac{1}{m!} + \frac{1}{N^m}\sum_{k=m+1}^\infty \frac{N^k}{k!}\right] \\
&= \frac{1}{N^{m+1}} \times \sum_{k=m+1}^\infty \frac{N^k}{k!}.
\end{aligned}$$

Use has been made of properties eq.(6.56), eq.(6.57) and the conjectured general formula eq.(6.58) is assumed true for value m. Its truth is thus seen to hold for $m + 1$ and the proof of our conjecture is complete.

Combining eq.(6.55) and eq.(6.58) yields

$$\begin{aligned}
\mathcal{N} &= \frac{(d-1)!}{N^{d-1}} \sum_{k=d-1}^\infty \frac{N^k}{k!} \\
&= \frac{(d-1)!}{N^{d-1}}\left[e^N - \sum_{k=0}^{d-2}\frac{N^k}{k!}\right]. \tag{6.59}
\end{aligned}$$

125

For later use we shall need the derivative of this with respect to N. This is

$$
\begin{aligned}
\frac{d}{dN}[\mathcal{N}] &= \frac{d}{dN}\left\{\frac{(d-1)!}{N^{d-1}}\left[e^N - \sum_{k=0}^{d-2}\frac{N^k}{k!}\right]\right\} \\
&= -\frac{(d-1)}{N}\mathcal{N} + \frac{(d-1)!}{N^{d-1}}\left[e^N - \sum_{k=0}^{d-3}\frac{N^k}{k!}\right] \\
&= -\frac{(d-1)}{N}\mathcal{N} + \mathcal{N} - \frac{(d-1)!}{N^{d-1}}\times\frac{N^{d-2}}{(d-2)!} \\
&= -\frac{(d-1)}{N}\mathcal{N} + \mathcal{N} - \frac{(d-1)}{N}.
\end{aligned}
\tag{6.60}
$$

From this we form the useful quantity

$$
\frac{N}{\mathcal{N}}\frac{d}{dN}[\mathcal{N}] = -(d-1) + N - \frac{(d-1)}{\mathcal{N}}.
\tag{6.61}
$$

Referring back to eq.(6.53) it is clear that the inferred distribution for ψ is given by

$$
p(\psi|\phi) = p(\phi|\psi).
$$

All inferred distributions have the same functional form and the same normalisation. It is then clear that the Correlation Information for the uniform measurement set is given by

$$
\{\psi,\phi\} = \int\int p(\psi|\phi)\log p(\psi|\phi)\,d\hat{\Omega}_{\vec{\psi}}d\hat{\Omega}_{\vec{\phi}}.
\tag{6.62}
$$

Substituting for $p(\psi|\phi) = p(\phi|\psi)$ from eq.(6.53) yields

$$
\begin{aligned}
\{\psi,\phi\} =\ &\int\int\frac{1}{\mathcal{N}}\exp N\{<\phi|\psi><\psi|\phi>\} \\
&\times\log\left[\frac{1}{\mathcal{N}}\exp N\{<\phi|\psi><\psi|\phi>\}\right]d\hat{\Omega}_{\vec{\psi}}d\hat{\Omega}_{\vec{\phi}}.
\end{aligned}
$$

Exploiting the constant nature of the normalisation, the definition of the normalisation, and the symmetric appearance of ψ and ϕ this can be reduced to

$$
\{\psi,\phi\} = \log\frac{1}{\mathcal{N}} + \frac{N}{\mathcal{N}}\frac{d}{dN}[\mathcal{N}].
\tag{6.63}
$$

126

Now making use of eq.(6.61) we obtain

$$\{\psi, \phi\} = \log \frac{1}{\mathcal{N}} - (d-1) + N - \frac{(d-1)}{\mathcal{N}}. \qquad (6.64)$$

Recalling eq.(6.59) we can see that for large N the last term goes to zero whilst the first term can be reduced as follows:

$$
\begin{aligned}
\log \frac{1}{\mathcal{N}} &= \log \left[\frac{N^{d-1}}{(d-1)!} \right] - \log \left[e^N - \sum_{k=0}^{d-2} \frac{N^k}{k!} \right] \\
&= \log \left[\frac{N^{d-1}}{(d-1)!} \right] - \log[e^N] - \log \left[1 - e^{-N} \sum_{k=0}^{d-2} \frac{N^k}{k!} \right] \\
&\simeq \log \left[\frac{N^{d-1}}{(d-1)!} \right] - N.
\end{aligned}
$$

Asymptotically the Correlation Information is given by

$$\{\psi, \phi\} = \log \left[\frac{N^{d-1}}{(d-1)!} \right] - (d-1). \qquad (6.65)$$

Thus the conjecture made at the start of this section is confirmed. Also note that substitution of $d = 2$ into the above result recovers that found for two–state systems in §5.8, page 95.

6.7 Upper bound to the Correlation Information

The general argument is very similar to that in §5.9. The corresponding output of greatest information is the all yes singlet result. This has the inferred distribution

$$p(\psi | \Phi_N) = \frac{1}{\mathcal{N}} \left(<\psi|\phi><\phi|\psi> \right)^N. \qquad (6.66)$$

Such a function scales as $k = N$ according to the definition given in eq.(6.1). Application of eq.(6.6) shows that the normalisation is

$$\mathcal{N} = \frac{(d-1)! N!}{(d+N-1)!}. \qquad (6.67)$$

127

If we pick up the argument of §5.9 at eq.(5.103) on page 103 one finds

$$\{\psi, \Phi_N\} \leq \int \frac{1}{\mathcal{N}} (x^2 + y^2)^N \log \left[\frac{1}{\mathcal{N}} (x^2 + y^2)^N\right] d\hat{\Omega}_{\hat{z}}. \qquad (6.68)$$

Here we have chosen to a basis such that

$$<\psi|\phi><\phi|\psi> = x^2 + y^2.$$

Now eq.(6.68) simplifies readily to

$$\{\psi, \Phi_N\} \leq \log \frac{1}{\mathcal{N}} + \frac{N}{\mathcal{N}} \int (x^2 + y^2)^N \log(x^2 + y^2) \, d\hat{\Omega}_{\hat{z}}. \qquad (6.69)$$

Use has been made of normalisation and the fact that $\mathcal{N}$ may be taken outside the integral. Given our experience with integrals of this type we may immediately write

$$
\begin{aligned}
\int (x^2 + y^2)^N \log(x^2 + y^2) \, d\hat{\Omega}_{\hat{z}} &= (d-1)! \int_0^\infty F_{d-1}(u) u^N \log u \, du \\
&= (d-1)! \int_0^\infty \delta(1-u)[u^N \log u]^{d-1} \, du.
\end{aligned}
$$

A little experimentation suggests the guess

$$
\begin{aligned}
[u^N \log u]^m &= \frac{N!}{(N+m)!} u^{N+m} \log u \\
&\quad - \left(\frac{1}{N+1} + \cdots + \frac{1}{N+m}\right) \frac{N!}{(N+m)!} u^{N+m}. \qquad (6.70)
\end{aligned}
$$

Verifying this for $m = 1$,

$$[u^N \log u] = \frac{u^{N+1}}{N+1} \log u - \frac{u^{N+1}}{(N+1)^2},$$

as required. Operating upon eq.(6.70) we have

$$
\begin{aligned}
[[u^N \log u]^m] &= \frac{N!}{(N+m)!} [u^{N+m} \log u] \\
&\quad - \left(\frac{1}{N+1} + \cdots + \frac{1}{N+m}\right) \frac{N!}{(N+m)!} [u^{N+m}]. \qquad (6.71)
\end{aligned}
$$

128

Using,

$$[u^{N+m} \log u] = \frac{u^{N+m+1}}{(N+m+1)} \left\{ \log u - \frac{1}{(N+m+1)} \right\}$$

and

$$\frac{1}{(N+m)!} [u^{N+m}] = \frac{u^{N+m+1}}{(N+m+1)!},$$

it is not difficult to see that the right–hand side of the expression eq.(6.71) is indeed the conjectured formula at $m+1$. Thus the proof is complete and we need only evaluate

$$[u^N \log u]^{d-1},$$

at $u = 1$. This yields from eq.(6.70) the result

$$\int_0^\infty \delta(1-u)[u^N \log u]^{d-1} \, du = - \left(\frac{1}{N+1} + \cdots + \frac{1}{N+d-1} \right) \frac{N!}{(N+d-1)!}.$$

Let

$$D = \left(\frac{1}{N+1} + \cdots + \frac{1}{N+d-1} \right)$$

and substitute for both terms in eq.(6.69) to find

$$\{\psi, \Phi_N\} \le \log \left[\frac{(d+N-1)!}{N!(d-1)!} \right] - N \times D. \qquad (6.72)$$

At $d = 2$, we recover the exact expression found in eq.(5.103), on page 103. For N large compared with d, one finds

$$N \times D \simeq d - 1.$$

Also,

$$\frac{(d+N-1)!}{N!(d-1)!} = \frac{N^{d-1}}{(d-1)!} \prod_{k=1}^{d-1} \left(1 + \frac{k}{N} \right);$$

so that

$$\log \left[\frac{(d+N-1)!}{N!(d-1)!} \right] \simeq \log \left[\frac{N^{d-1}}{(d-1)!} \right]. \qquad (6.73)$$

Thus the asymptotic upper bound for the Correlation Information is given by

$$\{\psi, \Phi_N\} \le \log \left[\frac{N^{d-1}}{(d-1)!} \right] - (d-1). \qquad (6.74)$$

129

Comparing this to eq.(6.65), we can see that the uniform measurement set achieves this bound and so corresponds to the optimal apparatus on Hilbert spaces of arbitrary finite dimensionality.

Chapter 7

Conclusion

7.1 Conclusions about Quantum Inference

We believe that we have demonstrated that the problem of empirical state determination can be usefully reformulated in a Bayesian context. The success of this program is largely due to the peculiar symmetry aspects of quantum theory. By this we mean that two of the major objections to Bayesian Inference, concerning the choice of prior and the subjective nature of the hypothesised correlation, can be dealt with on the basis of symmetry arguments.

In this respect, quantum measurement is naturally suited to such a treatment. The correlation that is involved in the description of any quantum observation process has universal form. Furthermore, the question of choosing a prior distribution over states can be disposed of through seeking a unitary–invariant measure. This is uniquely selected.

We have seen, explicitly in the two–dimensional case, that the usual notion of expectation values is recovered for large numbers of observations. Emphasis upon the inferred distribution over states makes it clear that there is no limit in principle to the precision with which the state of a finite-dimensional ensemble may be known. This may mean perfect knowledge of the matrix elements of an impure state. If the ensemble state is pure, then our inference will be an impure state for finite numbers of observations. In the limit of large N this approaches the correct pure state.

We have not addressed the question of measurability of the set of bases necessary for this program. The lack of a sufficient set may defeat complete determination of an ensemble state in practice. Clearly, the nature of such restrictions will depend upon the particular type of system involved. For complete generality the theory has been formulated solely in terms of the Hilbert space dimensionality. Most systems with a finite number of states are spin or angular momentum systems. We have avoided using this description in order to arrive at the most general problem. One would expect there to be restrictions upon particularising to this class of system.

The inclusion of concepts from Information Theory enabled us to establish a quantitative measure of the limits to information gain in quantum observation processes. In general we found the Correlation Information to be constrained asymptotically by the expression

$$\{\psi, \Phi_N\} \leq \log \left[\frac{N^{d-1}}{(d-1)!} \right] - (d-1), \tag{7.1}$$

where d is the Hilbert space dimensionality. We believe this to be a new result.

Also we discovered the interesting result:

$$p(\phi|\psi) = \frac{1}{N} \times \exp N\{<\phi|\psi><\psi|\phi>\}, \tag{7.2}$$

where this represents the effective N-trial correlation between the apparatus indication and the true state for the generalised uniform measurement set. The significance of this is unclear, but it is certainly unusual and worthy of some attention.

The uniform measurement set is a theoretical idealisation. Its usefulness lies in demonstrating that, in principle one may imagine a quantum state determination apparatus whose state space of readings is isomorphic to that of the system observed. This ties quantum observation to the theory of classical measurement. As a tool for thought, one can then consider states to be the primary observable with all other quantities derived by calculation. The only essential difference between the two is the fact that ensemble measurement is necessary. The analogy with communication systems is a strong one because we can now understand that many finite-state subsystems need to be observed in order to acquire a certain amount of information about the preparator state. Infinite information is associated with perfect knowledge of any variable and finite state systems function like a finite signal alphabet on what is essentially a discrete communication channel.

During the course of writing, the work of Balian[5, 6, 7] came to our notice. This concerns the problem of assigning a density matrix that is consistent with certain expectation value data. Use is made of the maximum entropy principle to make this assignment. The authors express some concern about the validity of this procedure. Now, it is well known that Gaussian distributions are the maximum entropy distributions consistent with a certain mean and variance. Given that we are able to construct an inferred distribution for the system state, which is Gaussian in virtue of the properties of the product correlation and reproduces the necessary projector expectation values, it may be that the two lines of approach can be connected. That is to say, the inherently Gaussian nature of inferred distributions may validate the maximum entropy procedure. If so, we believe that the Bayesian approach offers a better prospect for rigorous derivation, although we accept that this is a contentious issue.

One particularly attractive feature of this theory of state inference is that it makes obvious connection with some of the attempted pre–axiomatisations of quantum mechanics. We are refering to the work of Jauch et al.[4](The Geneva School) on lattices of propositions as a logical basis for the theory. A state is then defined to be a collection of yes–no propositions, assigned certain probabilities. Dynamics on this structure involves automorphisms of the lattice. The Geneva School discovered an interesting connection with Projective Geometry which motivated a realisation of these lattices upon vector spaces over the fields of Reals, Complex numbers or Quaternions. There ought to be generalisations of our theory to the other two cases of evolving probability lattices. Perhaps these could be interesting outside of quantum mechanics. We are indicating possible use of the quantum mechanical paradigm for any situation where the evolution of probability is required.[1]

The theory of Quantum Inference realises the empirical assignment of a state to an ensemble precisely in terms of measured probabilities. Such a procedure involves only the counting of alternatives manifested for each setting of the analyser. Such a minimal reliance upon extraneous concepts is attractive as a foundation for measurement. However, one must remember that we have given no indication of how the basis realised by a particular analyser setting is to be known. This is a serious drawback, although it is

---

[1]The required generalised Schrödinger equations would be obtained by stipulating invariance of the square–norm of states. We cannot offer a good suggestion along these lines. It was motivated by thinking about magnetised dice.

shared by more traditional approaches.

Finally, it would be worthwhile to attempt extension of these techniques to infinite dimensional Hilbert space. One would not expect this to be practically useful. However, it may provide an alternative route to developing an Information Theoretic Uncertainty Principle[45]. The central difficulty is the prescription of the nature of measurement bases. One might expect the coherent states[54] to be useful in this regard.

7.2   Tests of Transformation Theory

In this section we entertain some speculation upon quantum theory that is motivated by formalism. Then we turn to the one question of substance that emerges concerning the possiblity of testing quantum *evolution* within an operational methodology.

One of the most interesting formal features of non-relativistic quantum mechanics is that the Schrödinger equation can be motivated purely upon symmetry grounds, (Wigner's theorem). By this we mean that if it be stipulated that the space of states is a Hilbert space and one is interested in continuous evolution, then the equation:

$$i\hbar \frac{d}{dt}|\psi(t)> = \hat{H}(t)|\psi(t)>$$

is a necessary result. We are not concerned with the fact that $\hat{H}(t)$ is time dependent and that the equation may be very difficult to integrate. We simply stress that a time dependent evolution operator is so determined.[2]

Now it is often said that quantum mechanics cannot be formulated without reference to classical mechanics. This is not strictly true. It certainly can be, we need only specify a Hilbert space and an arbitrary time dependent $\hat{U}(t)$ to obtain a possible quantum dynamics. Of course this is not the sense in which such claims are made. Nevertheless, the reasoning employed is worthy of some attention.

The central difficulty is that such a statement is empty in the sense of any useful physics. We require guidance about how to select the space of states and the $\hat{U}(t)$. It is a curious feature of physics that natural laws are

---

[2]This should be so for finite dimensional space, from thinking about motion of a point on the Generalised Poincaré sphere. Perhaps not for infinite dimensional Hilbert space.

almost always of differential form (a form of locality). We are talking about finding global solutions, which is very uṋatural in classical physics because of the difficulty of proving global existence–uniqueness theorems for the general differential equations involved[26]. There would not appear to be such a difficulty with purely formal quantum mechanics.

Now we come to an interesting point. Quantum mechanics has made connection with classical physics through the special properties of the Heisenberg algebra.[3] That is to say, one chooses a representation and writes the Schrödinger equation in terms of the basis labels of *two* complete sets

$$|p> \quad \text{and} \quad |q>,$$

or rather a single one plus an operator for the second

$$\hat{p} = -i\hbar \frac{d}{dq}.$$

These orthonormal sets have a permanent and constant *skewness* expressed by an unchanging unitary operator:

$$<\mathbf{q}|\mathbf{p}> = (2\pi\hbar)^{-3/2} \times e^{i(\mathbf{p}\cdot\mathbf{q})/\hbar}.$$

These are skewed bases on an infinite–dimensional Hilbert space in the same sense as the mutually unbiased bases used in this work. They can indeed be related to these as a limit over prime–dimensional Hilbert spaces[27, p.259]. Planck's constant then appears as the reciprocal of a large prime number.

Such a device makes natural connection with classical physics through enabling two things:

- Quantisation, where it is possible to guess $\hat{H}(t)$ from knowledge of classical mechanics[4] replacing coordinates $(\mathbf{q}, \mathbf{p})$ with operators $(\hat{\mathbf{q}}, \hat{\mathbf{p}})$.

- A transformation from the space of evolving quantum states to a space of evolving classical states. This is done via the Wigner function and its Husimi variants.

---

[3]The canonical commutation relations: $[q, p] = i\hbar$.

[4]The *Stone–von Neumann* theorem is important in this regard. It links the constant $\hbar$ to the uniqueness of this procedure, see [53, p.137].

If it were ever possible to remove the bedrock of classical mechanics and the route it offers to quantisation; it would surely be necessary to derive it in some sense from quantum mechanics. We cite one interesting fact that has to do with the way in which a quantum Hamiltonian determines a classical Hamiltonian. This suggests the possibilty that if we can obtain $\hat{U}(t)$ in a way which determines $\hat{H}(t)$ then there is an avenue towards this goal.

The idea is to seek a Hamiltonian–less quantum mechanics.[5] An initial model might involve searching for a natural connection between two simple finite–dimensional systems so that they co-evolve in some uniquely selected fashion. There is no evidence that this should be a fruitful line to follow. However, it is intuitively attractive because, against wisdom, we may suggest that this ought to be the case in that a Hamiltonian in some sense stands for *the rest of the Universe*, which if we were to apply the theory in totality must also be a state. There would be obvious problems incorporating time into such a theory.

Even if it were possible to carry out such a program then it would still be necessary to explain *wave packet collapse*, that is to understand Process One within a supposedly deterministic theory. Everett's ideas are frequently raised in this connection. Ultimately the point that must be understood is how one acquires knowledge in a mutually disturbing world.

The traditional stipulation that measurement apparatus be *classical* is in essence a statement that we know for sure what a piece of apparatus, or any physical stuff, is doing. This leads to the probabilistic interpretation of Born, which is a way of deriving certain information in a statistical fashion. The present work shows that one can reformulate things only slightly to turn this into a natural theory for finding actual states. Quantum mechanics then limits knowledge about the state of particular systems, but does not prohibit certainty about the preparation if the observational procedure can be repeated with an identical input state arbitrarily many times. This limitation to ensemble measurement only should come as no suprise since the revision is one towards interconnection, the inseparability of system and apparatus.

Of course the above is pure speculation and so not very useful. However, having mentioned the possibility of theories that seek $\hat{U}(t)$, it is worthwhile to explore the avenues available for making an empirical assignment.

Without much further effort it is possible to turn the theory of Quantum

---

[5]Mention of this is made in Klauder[54, p.167].

State Inference into one where the goal is to characterise a region of space by its evolution operator. For example, take the standard Stern–Gerlach experiment. With this apparatus it is certainly possible to realise an analyser-preparator couple. The empirical observation of the beam splitting into two parts finds explanation in terms of spin–half particles and their two–state property. With this alone in mind we can certainly attempt to assign a state to the preparator by using beam–branching intensity ratio measurements for three different orientations of the analyser.

Doing this for several different orientations of the preparing magnet allows testing of the transformation rule for space rotations. We could apply the state inference technique to test the predictions of quantum mechanics concerning the density matrix corresponding to a particular rotation. However, in this case such a test is much more easily carried out by simply measuring the change in beam–branching ratio according to the rule

$$\frac{1}{2}(1+\hat{\mathbf{s}} \cdot \hat{\mathbf{r}}) = \cos^2(\theta/2).$$

There is then no need to determine any states.

Something more useful is gained if we consider $\hat{U}$ that arise not from a rotation in space, which amounts to a change of representation, but due to some intervening dynamics between preparator and analyser. We may assert the same transformation equation

$$\rho = \hat{U} \, \rho_0 \, \hat{U}^\dagger,$$

which applies for any Process Two evolution and attempt, for a given situation, to infer the matrix $\hat{U}$. Observing ρ for many possible initial ρ_0 enables determination of the $\hat{U}$ that empirically characterises insertion of some *general evolver* into the beam path between the preparator and analyser.

To do this one requires a set of preparators that span the space of $\hat{U}$ and whose states are known through prior calibration with an analyser, see figure(7.1). Inserting an arbitrary *evolution*, some field or otherwise, between the preparator and analyser, the task is to accumulate a body of data in the form of inferred states for each of the preparators with and without the evolution in–between. The problem is one involving inference of the matrix $\hat{U}$[6] in terms of data:

$$\rho(i) = \hat{U} \, \rho_0(i) \, \hat{U}^\dagger,$$

---

[6]This is a kind of empirically assigned S–matrix.

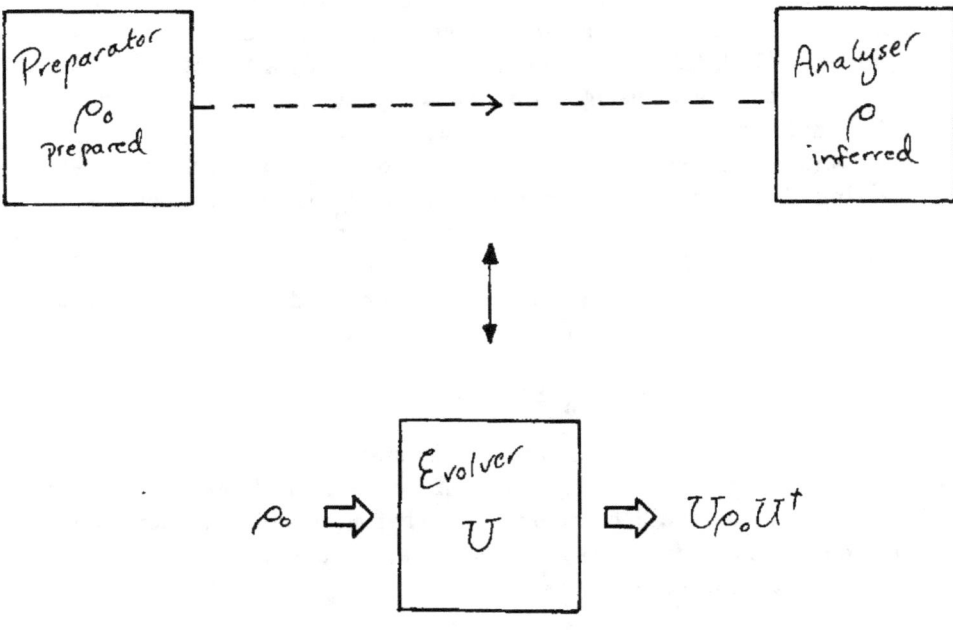

Figure 7.1: Schematic of proposed experimental set-up for empirical assignment of evolution operators. Universal preparator and analyser modules are required. After calibration of the preparators, a particular evolver is inserted into the beam path and the new preparator states are determined. Of interest are the evolution operator and possible changes in beam entropy.

138

where $\rho_0(i)$ denotes the initial inferred state for a given preparator and $\rho(i)$ denotes the new one after we switch on the evolution. Like the state determination problem, upon which such a procedure is based, there will be a minimal set of such states required to accumulate all of the necessary data. There will also be a confidence attached according to the number of observations made. It would be interesting to explore this question and see if it admits a similar Bayesian treatment.

Supposing that the above procedure can be implemented then one possibilty that is immediately striking involves measuring any changes in entropy that may result from partial collapse of an ensemble. This might be expected if Process One is a physical one with some governing time–scale. If this were to be observed in a particular instance then varying the time–of–flight through the evolution region would enable testing of this hypothesis. Of course it is also possible in principle to test various calculated evolutions.[7]

Further work is required on this; both in solving the evolution operator inference problem and in establishing useful tests of quantum theory. However, the mere ability to determine states is a powerful one and more direct than the usual tests involving the measurment of energy levels. It is suprising that not much attention has been given to this before although the possibilty of in–principle determination has been known for a long time. We believe that the Bayesian inference approach places the problem in a more natural setting, that should widen the possibilites for experiment.

---

[7]This may have technological significance in the construction of Quantum Computers. In such devices the process of computation involves the evolution of input states by a unitary operator with the program output being the result of a measurement performed upon the computer. They have interesting properties due to *quantum parallelism* which enables them to surpass limits upon classical computers imposed by the Church–Turing principle. See Deutsch[55, 56].

Appendix A

Asymptotic Information for Singlet measurement

The Correlation Information for Singlet measurement is given exactly by

$$\{\hat{\mathbf{r}}, \Phi_N\} = \log(N+1) - \frac{N}{2} + \frac{1}{N+1} \sum_{k=0}^{N} \log \binom{N}{k}. \qquad (A.1)$$

We wish to develop an asymptotic form. Therefore concentrate on the last term. The summands for $k = 0$ and $k = N$ do not contribute. The remainder can be expanded as follows:

$$
\begin{aligned}
\frac{1}{N+1} \sum_{k=1}^{N-1} \log \binom{N}{k} &= \frac{1}{N+1} \sum_{k=1}^{N-1} [\log N! - \log k! - \log(N-k)!] \\
&= \frac{N-1}{N+1} \log N! - \frac{2}{N+1} \sum_{k=1}^{N-1} \log k! \\
&= \log N! - \frac{2}{N+1} \sum_{k=1}^{N} \log k!. \qquad (A.2)
\end{aligned}
$$

Observe,

$$
\begin{aligned}
\frac{2}{N+1} \sum_{k=1}^{N} \log k! &= \frac{2}{N+1} \sum_{k=1}^{N} (N+1-k) \log k \\
&= 2 \log N! - \frac{2}{N+1} \sum_{k=1}^{N} k \log k. \qquad (A.3)
\end{aligned}
$$

A-1

Therefore,

$$\frac{1}{N+1}\sum_{k=1}^{N-1}\log\binom{N}{k} = \frac{2}{N+1}\sum_{k=1}^{N}k\log k \ - \ \log N!. \qquad (A.4)$$

Using Stirling's approximation:

$$\log N! = (N+1/2)\log N - N + \log\sqrt{2\pi} + \Theta_1,$$

where $0 < \Theta_1 < 1/12N$, the first term of eq.(A.4) can be massaged as follows.

$$\frac{2}{N+1}\sum_{k=1}^{N}k\log k \ = \ \frac{2}{N+1}\log N\sum_{k=1}^{N}k \ + \ \frac{2N^2}{N+1}\sum_{k=1}^{N}(1/N)(k/N)\log(k/N)$$

$$= \ N\log N \ + \ \frac{2N^2}{N+1}\times S_N. \qquad (A.5)$$

Here,

$$S_N = \sum_{k=1}^{N}(1/N)(k/N)\log(k/N)$$

and clearly satisifies

$$\int_0^1 u\log u\,du < \ S_N \ < \int_{1/N}^1 u\log u\,du. \qquad (A.6)$$

Evaluating upper and lower bounds,

$$-1/4 < \ S_N \ < -1/4 + \Theta_2,$$

where

$$\Theta_2 = \frac{1}{2N^2}(1/2 + \log N).$$

Substitution of S_N back in eq.(A.5) shows that

$$\frac{2}{N+1}\sum_{k=1}^{N}k\log k \ = \ N\log N - \frac{N^2}{2(N+1)} + \frac{2N^2}{N+1}\Theta_2. \qquad (A.7)$$

But,

$$\frac{2N^2}{N+1}\Theta_2 \ = \ \frac{1}{N+1}(1/2 + \log N),$$

A-2

which goes to zero as $N \to \infty$. To obtain the asymptotic formula both this and the error in Stirling's formula may be dropped. Combining eq.(A.4) with eq.(A.7) yields the following approximation to eq.(A.2),

$$\frac{1}{N+1} \sum_{k=1}^{N-1} \log \binom{N}{k} \sim N \log N - \frac{N^2}{2(N+1)}$$
$$- (N+1/2) \log N + N - \log \sqrt{2\pi}.$$

Notice that,

$$N - \frac{N^2}{2(N+1)} = \frac{N^2 + 2N}{2(N+1)} \sim \frac{N+1}{2},$$

so that the previous equation simplifies to

$$\frac{1}{N+1} \sum_{k=1}^{N-1} \log \binom{N}{k} \sim \frac{N+1}{2} - \log \sqrt{2N\pi}. \qquad (A.8)$$

Substitution of this in eq.(A.1) gives for the asymptotic Correlation Information

$$\{\hat{r}, \Phi_N\} \sim \log(N+1) - \frac{N}{2} + \frac{N+1}{2} - \log \sqrt{2N\pi}.$$

Discarding terms that go to zero for large N this can be rewritten as

$$\{\hat{r}, \Phi_N\} \sim \frac{1}{2} \log N - \frac{1}{2}(\log 2\pi - 1). \qquad (A.9)$$

This then is the required asymptotic form of the Correlation Information for Singlet measurement. Of course it is also the Correlation Information for the coin–tossing experiment.

Appendix B

Best Geometries for small N

Here we tabulate the coordinate directions for the best geometries found for $N \leq 10$ as listed in table(5.1). The interesting ones are the irregular geometries favoured by A_5, A_7 and A_8. Coordinate directions for the Platonic solids are included for completeness. Twelve significant figures were used in calculation.

Irregular Quintet		
$\hat{\mathbf{x}}$	$\hat{\mathbf{y}}$	$\hat{\mathbf{z}}$
0.000 000 000	0.000 000 000	1.000 000 000
0.915 763 507	0.000 000 000	−.401 717 811
−.915 763 507	0.000 000 000	−.401 717 811
0.000 000 000	0.915 763 507	−.401 717 811
0.000 000 000	−.915 763 507	−.401 717 811

Table B.1: Irregular geometry favoured by A_5.

Irregular Septet			
$\hat{x}$	$\hat{y}$	$\hat{z}$	reps.
0.000 000 000	0.000 000 000	1.000 000 000	1
−.440 114 116	0.762 299 997	−.474 550 585	2
−.440 114 116	−.762 299 997	−.474 550 585	2
0.880 228 234	0.000 000 000	−.474 550 585	2

Table B.2: Irregular geometry favoured by A_7. Three vectors are doubled up.

Irregular Octet			
$\hat{x}$	$\hat{y}$	$\hat{z}$	reps.
0.000 000 000	0.000 000 000	1.000 000 000	2
0.997 785 061	0.000 000 000	−.066 520 457	2
0.200 871 174	0.959 065 002	−.199 612 356	2
−.143 320 298	−.638 341 612	−.756 293 117	1
−.756 293 117	0.638 341 612	−.143 320 298	1

Table B.3: Irregular geometry favoured by A_8. Three vectors are doubled up.

Tetrahedron		
$\hat{x}$	$\hat{y}$	$\hat{z}$
0.000 000 000	0.000 000 000	1.000 000 000
0.942 809 025	0.000 000 000	−.333 333 333
−.471 404 513	0.816 496 611	−.333 333 333
−.471 404 513	−.816 496 611	−.333 333 333

Table B.4: Vertex directions for the tetrahedron.

Dodecahedron		
$\hat{x}$	$\hat{y}$	$\hat{z}$
0.000 000 000	0.000 000 000	1.000 000 000
0.525 731 095	0.723 606 796	0.447 213 590
0.850 650 795	−.276 393 206	0.447 213 590
0.000 000 000	−.894 427 180	0.447 213 590
−.850 650 795	−.276 393 206	0.447 213 590
−.525 731 095	0.723 606 796	0.447 213 590

Table B.5: Face directions for the dodecahedron. There are twelve faces but only six are useful since the others are inversions of these.

Icosahedron		
$\hat{x}$	$\hat{y}$	$\hat{z}$
0.000 000 000	0.607 062 000	0.794 654 471
0.577 350 263	0.187 592 474	0.794 654 471
0.356 822 087	−.491 123 474	0.794 654 471
−.356 822 087	−.491 123 474	0.794 654 471
−.577 350 263	0.187 592 474	0.794 654 471
0.000 000 000	0.982 246 947	0.187 592 473
0.934 172 357	0.303 531 004	0.187 592 473
0.577 350 263	−.794 654 471	0.187 592 473
−.577 350 263	−.794 654 471	0.187 592 473
−.934 172 357	0.303 531 004	0.187 592 473

Table B.6: Face directions for the icosahedron. There are twenty faces but only ten are useful because of inversion symmetry.

Bibliography

[1] Wootters, W.K. (1980), 'The Acquisition of Information from Quantum Measurements', Ph.D. thesis, (Cen. Th. Phys. Uni. of Texas, Austin).

[2] Wootters, W.K. and Zurek, W.H. (1982), 'A single quantum cannot be cloned', *Nature* **299**, p.802.

[3] Ghirardi, G.C. and Weber, T. (1983), 'Quantum Mechanics and Faster-Than-Light Communication Methodological Considerations', *Nuovo Cim.* **78B**, p.9.

[4] Jauch, J. M. (1968), *Foundations of Quantum Mechanics*, (Addison-Wesley).

[5] Balian, R. and Veneroni, M. (1987), 'Incomplete Descriptions, Relevant Information and Entropy Production in Collision Processes', *Ann. Phys.* **174**, p.229.

[6] Balian, R. and Balazs, N.L. (1987), 'Equiprobability, Inference, and Entropy in Quantum Theory', *Ann. Phys.* **179**, p.97.

[7] Balian, R. (1989), 'Gain of Information in a quantum measurement', *Eur. J. Phys.* **10**, p.208.

[8] Wheeler, J.A. and Zurek, W.H. (1983), *Quantum Theory and Measurement*, (Princeton).

[9] Jammer, M. (1974), *The Philosophy of Quantum Mechanics*, (Wiley).

[10] D'Espagnat, B. (1971), *Foundations of Quantum Mechanics*, (Academic).

[11] Penrose, R. (1989), *The Emperor's New Mind*, (Oxford).

[12] Prugovečki, E. (1971), *Quantum Mechanics in Hilbert Space*, (Academic Press).

[13] Gottfried, K. (1966), *Quantum Mechanics Vol. I: Fundamentals*, (Benjamin).

[14] Penrose, R. (1987), 'Newton Quantum Theory and Reality', in *300 Years of Gravitation*, eds. Hawking, S.W. and Israel, W. (Cambridge), p.17.

[15] D'Espagnat, B. (1989), *Reality and the Physicist*, (Cambridge).

[16] Rosen, J. (1982), *Symmetries in Physics: Selected Reprints*, (AAPT).

[17] Dirac, P.A.M. (1958), *The Principles of Quantum Mechanics*, 4th edn. (Oxford).

[18] Klauder, J.R. and Sudarshan, E.C.G. (1968), *Fundamentals of Quantum Optics*, (Benjamin).

[19] Born, M. and Wolf, E. (1980), *Principles of Optics*, 6th edn. (Pergamon).

[20] Bell, J.S. (1988), *Speakable and unspeakable in quantum mechanics*, (Cambridge).

[21] Deutsch, D. 'Quantum Communication thwarts eavesdroppers', *New Scientist* **1694**,p.25.

[22] Band, W. and Park, J.L. (1970), 'The Empirical Determination of Quantum States', *Found. Phys.* **1**, p.133.

[23] Band, W. and Park, J.L. (1971), 'A General Theory of Empirical State Determination in Quantum Mechanics: Part I', *Found. Phys.* **1**, p.211.

[24] Band, W. and Park, J.L. (1971), 'A General Method of Empirical State Determination in Quantum Mechanics: Part II', *Found. Phys.* **1**, p.339.

[25] Band, W. and Park, J.L. (1979), 'Quantum State Determination: Quorum for a particle in one dimension', *Am. J. Phys.* **47**, p.188.

[26] Mackey, G.W. (1963), *Mathematical Foundations of Quantum Mechanics*, (Benjamin).

[27] Schwinger, J. (1969), *Quantum Kinematics and Dynamics*, (Benjamin).

[28] Harriman, J.E. (1978), 'Geometry of density matrices ...', *Phys. Rev.* **A17**, p.1249.

[29] Wooters, W.K. (1985), 'Quantum Mechanics without Probabilty Amplitudes', in *Between Quantum and Cosmos*, eds. Zurek, et al. (Princeton), p.507.

[30] Ivanovič, I.D. (1981), 'Geometrical description of quantal state determination', *J. Phys. A:Math. Gen.* **14**, p.3241.

[31] Korn, G.A. and Korn, T.M. (1968), *Math. Hand. for Sci. and Eng.*, 2nd edn. (McGraw–Hill).

[32] Wootters, W.K. (1986), 'The Discrete Wigner Function', *Ann. N.Y.A.Sci.* **480**, p.275.

[33] Gnedenko, B.V. (1968), *Theory of Probability*, 4th edn. (Chelsea).

[34] Von Mises, R. (1964), *Mathematical Theory of Probability and Statistics*, (Academic).

[35] Gradshteyn, I.S. and Ryzhik, I.M. (1983), *Table of Integrals, Series and Products.*, 4th edn. (Academic).

[36] Haken, H. (1977), *Synergetics*, (Springer).

[37] Braunstein, S.L. and Caves, C.M. (1988), 'Information Theoretic Bell Inequalities', *Phys. Rev. Lett.* **61**, p.662.

[38] Lindblad, G. (1972), 'An Entropy Inequality for Quantum Measurement', *Commun. Math. Phys.* **28**, p.245.

[39] Lindblad, G. (1973), 'Entropy, Information and Quantum Measurements', *Commun. Math. Phys.* **33**, p.305.

[40] Helstrom, C.W., Liu, J.W.S. and Gordon, J.P. (1970), 'Quantum-Mechanical Communication Theory', *Proc. IEEE* **58**, p.1578.

[41] Shannon, C.E. and Weaver, W. (1949), *The Mathematical Theory of Communication*, (Illinois Press).

[42] Guiaşu, S. (1977), *Information Theory with Applications*, (McGraw–Hill).

[43] Gallager, R.G. (1968), *Information Theory and Reliable Communication*, (Wiley).

[44] Brillouin, L. (1962), *Science and Information Theory*, (Academic).

[45] Everett III, H. (1957), 'The Theory of the Universal Wave Function', Ph.D. thesis, pub. in *The Many–Worlds Interpretation of Quantum Mechanics*, eds. De Witt, B. and Graham, N. (Princeton, 1973), p.3.

[46] Bender, C.M. and Orszag, S.A. (1978), *Adv. Math. Meth. for Sci. and Eng.*, (McGraw–Hill).

[47] Itzykson, C. and Zuber J.B. (1987), *Quantum Field Theory*, 3rd edn. (McGraw–Hill).

[48] Gudder, S.P. (1979), *Stochastic Methods in Quantum Mechanics*, (Nth. Holland).

[49] Press, W.H., Flannery, B.P., Teukolsky, S.A. and Vetterling, W.T. (1986), *Numerical Recipes–The Art of Scientific Computing*, (Cambridge).

[50] Penrose, R. (1971), 'Angular Momentum: an Approach to Combinatorial Space–Time', in *Quantum Theory and Beyond*, ed. Bastin T. (Cambridge), p.151.

[51] Goodyear, C.C. (1971), *Signals and Information*, (Butterworths).

[52] Penrose, R. and Rindler, W. (1986), *Spinors and space–time: Vol I*, (Cambridge)

[53] Hermann, R. (1966), *Lie Groups for Physicists*, (Benjamin).

[54] Klauder, J.R. and Skagerstam, B.S. (1985), *Coherent States: Applications in Physics and Mathematical Physics*, (World Scientific).

[55] Deutsch, D. (1985), 'Quantum theory, the Church–Turing principle and the universal quantum computer' *Proc. Roy. Soc.* **A400**,p.97.

[56] Deutsch, D. (1987), 'Quantum Computers', *Computer Bulletin* **6**, p.24.